RESEARCH OPPORTUNITIES FOR MANAGING THE DEPARTMENT OF ENERGY'S TRANSURANIC AND MIXED WASTES

Committee on Long-Term Research Needs for Managing Transuranic and Mixed Wastes at Department of Energy Sites

Board on Radioactive Waste Management

Division on Earth and Life Studies

NATIONAL RESEARCH COUNCIL
OF THE NATIONAL ACADEMIES

THE NATIONAL ACADEMIES PRESS
Washington, D.C.
www.nap.edu

THE NATIONAL ACADEMIES PRESS • 500 Fifth Street, N.W. • Washington, DC 20001

NOTICE: The project that is the subject of this report was approved by the Governing Board of the National Research Council, whose members are drawn from the councils of the National Academy of Sciences, the National Academy of Engineering, and the Institute of Medicine. The members of the committee responsible for the report were chosen for their special competences and with regard for appropriate balance.

Support for this study was provided by the U.S. Department of Energy under Grant No. DE-FC01-99EW59049. All opinions, findings, conclusions, or recommendations expressed herein are those of the authors and do not necessarily reflect the views of the U.S. Department of Energy.

International Standard Book Number 0-309-08471-7

Additional copies of this report are available from:

The National Academies Press
500 Fifth Street, N.W.
Box 285
Washington, DC 20055
800-624-6242
202-334-3313 (in the Washington Metropolitan Area)
http://www.nap.edu

COVER PHOTO. DOE's inventory of transuranic and mixed wastes is large and heterogeneous. Most is stored in 55-gallon drums or larger containers. Photograph courtesy of the U.S. Department of Energy.

Printed in the United States of America

THE NATIONAL ACADEMIES

Advisers to the Nation on Science, Engineering, and Medicine

The **National Academy of Sciences** is a private, nonprofit, self-perpetuating society of distinguished scholars engaged in scientific and engineering research, dedicated to the furtherance of science and technology and to their use for the general welfare. Upon the authority of the charter granted to it by the Congress in 1863, the Academy has a mandate that requires it to advise the federal government on scientific and technical matters. Dr. Bruce M. Alberts is president of the National Academy of Sciences.

The **National Academy of Engineering** was established in 1964, under the charter of the National Academy of Sciences, as a parallel organization of outstanding engineers. It is autonomous in its administration and in the selection of its members, sharing with the National Academy of Sciences the responsibility for advising the federal government. The National Academy of Engineering also sponsors engineering programs aimed at meeting national needs, encourages education and research, and recognizes the superior achievements of engineers. Dr. Wm. A. Wulf is president of the National Academy of Engineering.

The **Institute of Medicine** was established in 1970 by the National Academy of Sciences to secure the services of eminent members of appropriate professions in the examination of policy matters pertaining to the health of the public. The Institute acts under the responsibility given to the National Academy of Sciences by its congressional charter to be an adviser to the federal government and, upon its own initiative, to identify issues of medical care, research, and education. Dr. Harvey V. Fineberg is president of the Institute of Medicine.

The **National Research Council** was organized by the National Academy of Sciences in 1916 to associate the broad community of science and technology with the Academy's purposes of furthering knowledge and advising the federal government. Functioning in accordance with general policies determined by the Academy, the Council has become the principal operating agency of both the National Academy of Sciences and the National Academy of Engineering in providing services to the government, the public, and the scientific and engineering communities. The Council is administered jointly by both Academies and the Institute of Medicine. Dr. Bruce M. Alberts and Dr. Wm. A. Wulf are chair and vice chair, respectively, of the National Research Council.

www.national-academies.org

COMMITTEE ON LONG-TERM RESEARCH NEEDS FOR MANAGING TRANSURANIC AND MIXED WASTES AT DEPARTMENT OF ENERGY SITES

LLOYD A. DUSCHA, Chair, U.S. Army Corps of Engineers (Retired), Reston, Virginia
CAROL J. BURNS, Los Alamos National Laboratory, New Mexico
RICHARD J. COLTON, Naval Research Laboratory, Washington, D.C.
KIMBERLEE J. KEARFOTT, Nuclear Engineering and Radiological Sciences, University of Michigan, Ann Arbor
RICHARD J. SAMELSON, PPG Industries (Retired), Pittsburgh, Pennsylvania
ROBERT J. STEFFAN, Envirogen, Inc., Lawrenceville, New Jersey
VICTORIA J. TSCHINKEL, Environmental Policy and Management, Tallahassee, Florida
MARIA E. UHLE, University of Tennessee, Knoxville
GERBEN J. ZYLSTRA, Rutgers University, New Brunswick, New Jersey

STAFF

JOHN R. WILEY, Study Director
DARLA J. THOMPSON, Research Assistant
LATRICIA C. BAILEY, Senior Project Assistant

BOARD ON RADIOACTIVE WASTE MANAGEMENT

JOHN F. AHEARNE, Chair, Sigma Xi and Duke University, Research Triangle Park, North Carolina
CHARLES MCCOMBIE, Vice Chair, Consultant, Gipf-Oberfrick, Switzerland
ROBERT M. BERNERO, U.S. Nuclear Regulatory Commission (retired), Gaithersburg, Maryland
ROBERT J. BUDNITZ, Future Resources Associates, Inc., Berkeley, California
GREGORY R. CHOPPIN, Florida State University, Tallahassee
RODNEY EWING, University of Michigan, Ann Arbor
JAMES H. JOHNSON, JR., Howard University, Washington, D.C.
HOWARD C. KUNREUTHER, University of Pennsylvania, Philadelphia
NIKOLAY LAVEROV, Russian Academy of Sciences, Moscow
MILTON LEVENSON, Bechtel International (retired), Menlo Park, California
JANE C.S. LONG, Mackay School of Mines, University of Nevada, Reno
ALEXANDER MACLACHLAN, E.I. du Pont de Nemours and Company (retired), Wilmington, Delaware
NORINE E. NOONAN, College of Charleston, South Carolina
EUGENE A. ROSA, Washington State University, Pullman
ATSUYUKI SUZUKI, University of Tokyo, Japan
VICTORIA J. TSCHINKEL, Environmental Policy and Management, Tallahassee, Florida

STAFF

KEVIN D. CROWLEY, Director
MICAH D. LOWENTHAL, Staff Officer
BARBARA PASTINA, Senior Staff Officer
JOHN R. WILEY, Senior Staff Officer
TONI GREENLEAF, Administrative Associate
DARLA J. THOMPSON, Research Assistant
LATRICIA C. BAILEY, Senior Project Assistant
LAURA D. LLANOS, Senior Project Assistant
ANGELA R. TAYLOR, Senior Project Assistant
JAMES YATES, JR., Office Assistant

Preface

The production of nuclear materials for the national defense, beginning in the 1940s and continuing until the end of the Cold War, led to the accumulation of large quantities of radioactive wastes at sites throughout the country. Site cleanup is now a major, long-term task for the Department of Energy (DOE). Transuranic waste and mixed low-level waste are contaminated with relatively low amounts of actinide isotopes or fission products, respectively, and with hazardous chemicals. These wastes include such diverse materials as process residues, construction debris, equipment, and trash. Early on these wastes were buried in trenches and landfills or managed by the use of seepage and evaporation ponds. These practices were recognized as inadequate, and since 1970 these wastes have been stored for retrieval, mostly in 55-gallon drums (see cover photo).

The stored inventory totals about 155,000 cubic meters, the equivalent of about three-quarters of a million drums. At least some of the approximately 500,000 cubic meters of buried waste will be retrieved. Ongoing DOE site cleanup efforts, such as stabilizing highly radioactive tank wastes and decommissioning production facilities, will result in further accumulation of transuranic and mixed wastes. Transuranic waste, which makes up more than two-thirds of the stored inventory and nearly a third of the buried inventory, is destined for permanent disposal in the Waste Isolation Pilot Plant, in a deep-underground salt formation in New Mexico. Mixed low-level waste will be disposed in licensed near-surface facilities operated by private contractors, although some will be disposed at DOE sites.

To help reduce costs and accelerate the schedule of its overall site cleanup program, DOE is making a concerted effort to retrieve and dispose of transuranic and mixed wastes as rapidly as possible. However, work with these wastes is only beginning, and it will continue for at least 20 years. Many current procedures are cumbersome and expensive. For example, each 55-gallon drum, or other container, must be handled individually several times to determine its contents and prepare it for shipment and disposal. Any efficiencies or added effectiveness that can be gained in these procedures will reduce labor and potential risks to workers, lower costs, and accelerate the schedule. To enable such endeavors, basic research is considered a vital tool.

The Congress recognized the essentiality of research and in 1995 chartered the Environmental Management Science Program (EMSP) to bring the nation's scientific capability to bear on the difficult, long-term cleanup challenges facing DOE. To assist in this effort, the National Academies have been requested on several occasions to provide advice in developing a research agenda for the EMSP. To that end, this report is the result of a study by the National Research Council Committee on Long-Term Research Needs for Managing Transuranic and Mixed Wastes at Department of Energy Sites.

To launch the study, the committee heard presentations from headquarters personnel on the policy and programmatic aspects of the Environmental Management Science Program. During the course of its study, the committee visited five sites to witness ongoing work on characterization, treatment, shipping preparation, and disposition and held meetings to receive presentations from site DOE and contractor personnel, as well as stakeholders with an interest in DOE cleanup activities.

On behalf of the committee, I would like to thank DOE headquarters, field offices, sites, and laboratory staffs, as well as the contractors and many other individuals who provided information to be used in this study for their time, patience, and openness in sharing their views on research needs. The committee found many knowledgeable, informed, and concerned people in DOE and among the contractors; many of their ideas are reflected in the consensus recommendations of the committee. Information provided by members of the DOE Office of Science and Technology's Transuranic and Mixed Waste Focus Area was especially useful.

I also wish to thank and recognize the staff of the National Academies Board on Radioactive Waste Management for their willing, efficient, and most capable assistance during the study in guiding the committee through the fact-finding, report-writing, and review phases, as well as in handling the myriad of logistic details for the committee members.

Lastly, I want to deeply thank the members of the committee for their dedication and diligence. Although of diverse background, they respected the overall goal of the study and report, and each made significant contributions. It was a pleasure working with the committee members and the staff of the Board on Radioactive Waste Management.

Lloyd A. Duscha
Chair

List of Report Reviewers

This report has been reviewed in draft form by individuals chosen for their diverse perspectives and technical expertise, in accordance with procedures approved by the National Research Council (NRC) Report Review Committee. The purpose of this independent review is to provide candid and critical comments that will assist the institution in making the published report as sound as possible and to ensure that the report meets institutional standards for objectivity, evidence, and responsiveness to the study charge. The content of the review comments and draft manuscript remains confidential to protect the integrity of the deliberative process. We wish to thank the following individuals for their participation in the review of this report:

Hugh Davis, Environmental Protection Agency
Catherine Fenselau, University of Maryland
Alexander MacLachlan, E.I. du Pont de Nemours & Company (retired)
Norine Noonan, College of Charleston, South Carolina
Gary Phillips, Georgetown University Medical Center
Gary Sayler, University of Tennessee
Bruce Thomson, University of New Mexico

Although the reviewers listed above have provided many constructive comments and suggestions, they were not asked to endorse the conclusions or recommendations, nor did they see the final draft of the report before its release. The review of this report was overseen by Kent F. Hansen, Massachusetts Institute of Technology, appointed by the Division on Earth and Life Studies, who was responsible for making certain that an independent examination of this report was carried out in accordance with NRC procedures and that all review comments were carefully considered. Responsibility for the final content of this report rests entirely with the authoring committee and the NRC.

Contents

EXECUTIVE SUMMARY **1**

1 INTRODUCTION, BACKGROUND, AND TASK **10**
Statement of Task, 11

2 FRAMING DOE'S TRANSURANIC AND MIXED WASTE CHALLENGES **14**
DOE's Transuranic and Mixed Waste Inventory, 16
Current and Evolving Regulatory Constraints, 23
Public Concerns, 31
Summary: Meeting TM Waste Challenges, 33

3 RESEARCH NEEDS AND OPPORTUNITIES **35**
Characterization, 36
Retrieval of Buried Waste, 49
Treatment, 56
Long-Term Monitoring, 69
Near-Term and Longer-Term Research, 73

REFERENCES **75**

APPENDIXES
A Overview of the Environmental Management Science Program and Pending Changes **85**
B The Transuranic and Mixed Waste Inventory **91**
C History of Alternatives to Incineration **103**
D Biographical Sketches of Committee Members **107**
E Presentations to the Committee **113**
F List of Acronyms **117**

Executive Summary

The National Academies' National Research Council (NRC) undertook this study to provide advice to the Department of Energy's (DOE's) Environmental Management Science Program on a long-term research agenda for managing and disposing of transuranic and mixed wastes. DOE's inventory of transuranic and mixed wastes (TM wastes) includes about 155,000 cubic meters of waste stored on some 30 DOE sites and another 450,000 cubic meters of buried waste—at least some of which is likely to require retrieval in the course of DOE's site cleanup program. Most of the stored inventory is in 55-gallon drums or other containers.[1] Although some of the buried waste is similarly packaged, knowledge of the condition of the containers and their contents is limited.

While DOE is making a concerted effort to accelerate the removal of TM wastes from its sites, the size of the inventory translates to a multi-decade effort that will require handling, characterizing, shipping, and disposing of hundreds of thousands of waste drums and other containers at a total cost of billions of dollars. Thus, there are sufficient time and strong incentives—safety, cost, and efficiency—for research toward developing new technologies for managing DOE's TM wastes and improving the scientific basis for public and regulatory decision making.

Transuranic (TRU) wastes comprise a variety of waste materials (e.g., trash, equipment, soil, sludge) that are contaminated with plutonium or other transuranic isotopes. Mixed low-level waste (MLLW) is similar to TRU waste except it contains small amounts of radioactive fission products as well as substances designated as hazardous by the Environmental Protection Agency (EPA). TRU wastes are intended for disposal at the Waste Isolation Pilot Plant (WIPP), which is in a deep salt formation in

[1]One cubic meter is equivalent in volume to five 55-gallon (200-liter) drums, so the stored inventory amounts to about three-quarters of a million drums.

southeastern New Mexico. MLLW can be disposed in facilities at or near the earth's surface that are constructed in compliance with EPA and other applicable regulations.

In 1995, Congress chartered the Environmental Management Science Program (EMSP) to bring the nation's scientific capability to bear on the difficult, long-term cleanup challenges facing DOE. To fulfill its charter, the EMSP solicits proposals and selectively funds research on problems relevant to the needs of DOE's Office of Environmental Management (EM). The objective of this study is to provide recommendations to the EMSP for the development of a research agenda to address challenges in managing TM wastes that are currently stored at DOE sites or will be produced as part of DOE's site cleanup program.

When this study was in its closing phases, DOE's Office of Environmental Management completed a "top-to-bottom" review, which will result in significant changes within EM and its Office of Science and Technology (OST), the sponsor of this study, to be effective at the beginning of fiscal year 2003. The five OST focus areas—including the Transuranic and Mixed Waste Focus Area (TMFA)—around which OST had previously organized its research and development activities will be abolished and replaced by two science and technology "thrusts." The EMSP will be removed entirely from EM and placed in the Office of Biological and Environmental Research within DOE's Office of Science.

The committee[2] did not attempt to assess the effects that this reorganization will have on the EMSP. However, the committee did note that the TMFA provided much of the technology needs and development information used in preparing this report. Without the focus area structure it may be more difficult for the EMSP to identify site technology needs and, especially, to keep a perspective on long-term needs that can be addressed through scientific research. Maintaining the relevance of its funded research to site cleanup needs will be important for the EMSP after the reorganization is completed in fiscal year 2003—for example, by continuing the joint review of research proposals by both OST for relevance to EM's needs and the Office of Science for scientific merit (see Appendix A).

---

[2]The Committee on Long-Term Research Needs for Managing Transuranic and Mixed Wastes at Department of Energy Sites, which developed this report, is referred to as "the committee" throughout. The committee completed its work in May 2002, about five months before the reorganization was to be finalized.

Challenges

Radioactive waste materials began accumulating in the 1940s with the development of the atomic bomb and continued through the Cold War. Although DOE has halted its production activities, TM wastes continue to accumulate, albeit at a slower rate, mainly from site cleanup and deactivation and decommissioning activities.

The challenges in managing and disposing TM wastes are largely attributable to the following:

- a large and highly diverse waste inventory, which is incompletely characterized;
- complex and evolving regulatory constraints from various agencies; and
- public concern and often opposition to technologies that are unfamiliar or that might change agreed-upon cleanup plans.

These challenges will affect the priorities of any research agenda developed by the EMSP.

DOE's greatest technical challenges for managing and disposing of its TM wastes arise from the sheer size of the inventory—characterizing the contents of hundreds of thousands of waste containers, retrieving at least a portion of buried wastes, providing treatments as necessary, and shipping the wastes to designated disposal facilities. The number of regulatory agencies and myriad applicable rules can produce conflicting or excessive requirements that lead to delays and increase costs. DOE has begun seeking regulatory changes in several specific instances (see Chapter 2). Public opposition to incineration, the technology DOE intended to use to treat a large portion of its TM wastes, has forced DOE to seek alternatives.

From these challenges, two clear roles for EMSP research arise:

1. To provide the scientific basis for new technologies that will be necessary for improving management and disposal of TM wastes during the next 20 years, especially if regulatory changes that DOE expects to simplify dealing with problematic wastes are not forthcoming.
2. To enhance the scientific information available for regulatory decision making and public involvement, including evidence that disposal systems are operating as intended.

Findings and Research Recommendations

After visiting DOE sites, considering the views expressed by a wide range of participants, and conducting internal deliberations, the committee concluded that the most significant research needs and opportunities lie in

- waste characterization and how the waste characteristics may change with time,
- location and retrieval of buried wastes,
- waste treatment, and
- long-term monitoring.

The committee has been selective in its recommendations to encourage concentration of limited funding to a few specific areas believed to make the most significant contributions to meeting future waste challenges. The recommendations were developed from presentations to the committee, site needs, apparent knowledge gaps, the potential for future cost and schedule savings, and the possibility of achieving technological breakthroughs. These recommendations deliberately were cast to reflect the goals of the research rather than what research is to be done. The latter is better left to the ingenuity of the scientists who will submit EMSP proposals.

Characterization

The EMSP should support research to improve the efficiency of characterizing DOE's TRU and mixed waste inventory. This should include research toward developing faster and more sensitive characterization and analysis tools to reduce costs and accelerate throughput. It should also include research to develop a fuller understanding of how waste characteristics may change with time (chemical, biological, radiological, and physical processes) to aid in decision making about disposition paths and to simplify the demonstration of regulatory compliance.

Determining the physical, chemical, and radiological properties of TM wastes pertinent to handling, processing, transportation, and storage is costly and time-consuming. The problem is amplified by the wide variety of the wastes and their heterogeneity. Improving and simplifying waste characterization can reduce costs and increase the rate of shipping wastes to disposal facilities.

The committee found needs for faster and more sensitive characterization technologies, for making automated sampling more reliable, and for improving statistical sampling methods. There is a lack in basic knowledge of how waste characteristics may change with time, including both short-term changes that affect storage and shipment and long-term changes that may occur in a disposal facility. This lack of knowledge drives conservatism in characterization, transportation, and disposal requirements. Possible microbial effects in waste have generally been ignored.

The committee believes that the greatest challenges for the next generation of characterization technologies will be to provide the following:

- more rapid, automated nondestructive assay and evaluation methods;
- more sensitive nondestructive assay and evaluation technologies for larger containers and hard-to-detect contaminants; and
- improved methods, based on fundamental modeling, to derive present and future waste characteristics from a limited number of sampling parameters.

Research toward new, noninvasive, remote imaging and image recognition methods and in-drum sensors to provide faster and more sensitive technologies for characterization could lead to significant savings in time, cost, and risk of worker exposure. While noninvasive diagnostics are ideal, the use of minimally invasive sensors also has promise. Research on microbial activity in TM waste may lead to new ways to control long-term changes in waste stability or toxicity. One of the most beneficial cost-saving tools would be the formulation of more reliable predictive models, validated by experimental data, of how waste characteristics may change with time. This would be most useful in predicting deleterious processes that might occur in the waste, such as gas generation or matrix degradation.

Retrieval of Buried Waste

The EMSP should support research that will facilitate management of buried TRU and mixed waste in anticipation that retrieval of some waste will become necessary. This research should emphasize remote imaging and sensing technologies to locate and identify buried waste and retrieval methods that enhance worker safety.

Given the complex and changing nature of regulatory requirements and public perception, the committee believes that some buried wastes are likely to be retrieved in the future. Burial was largely in near-surface excavations—some wastes in containers and some in bulk.

The committee believes that the greatest challenges for the next generation of retrieval technologies will be to provide

- improved, noninvasive means to locate and identify buried waste whether or not it is containerized;
- remote, noninvasive assessment of the condition of waste containers and of potential leakage from the containers; and
- remote intelligent machines (robots) for waste retrieval and repackaging or treatment as necessary.

Before buried waste can be retrieved, it must be located and its condition determined. Determining the integrity of a waste container prior to retrieval extends the challenges of imaging science to objects below ground. In addition to improving image resolution, research is needed to improve identification of the object, the surrounding contamination, and the stability of the contaminants.

Intact drums could be retrieved and characterized using the processes developed for stored waste; however, it would be preferable to perform characterization at the burial site as each drum is retrieved to minimize handling and ensure worker safety. Robotic devices would help protect workers by handling containers that emit radiation or have been breached and have radioactive contamination on their surface or in the surrounding soil.

Microorganisms can have a profound impact on the chemistry and fate of buried waste. Although many biological studies have focused on a better understanding of the environmental fate of radioactive and toxic metals, few studies have investigated the complex relationships among microbes and the organic and inorganic constituents of TM waste. Understanding these relationships could lead to improved predictability of the long-term fate and risk of the waste materials.

Treatment

The EMSP should support research for treating TRU and mixed waste to facilitate disposal. This research should include processes to simplify or stabilize waste, with emphasis on improving metal separations, eliminating incinerator emissions, and enabling alternative organic destruction methods.

Treatment includes operations intended to improve the safety and/or economy of managing waste by changing the characteristics of the waste—volume reduction, removal of radionuclides or other contaminants, and altering the waste composition. Treatment is necessary if waste does not meet shipping requirements or acceptance criteria at the intended disposal site.

In the absence of effective treatment technologies, waste is simply repackaged to avoid the problem. Repackaging waste in order to meet shipping requirements is extremely inefficient, may increase volume manyfold, and presents hazards to workers. Phasing out incineration for the destruction of organic constituents requires the development of alternative technologies. Wastes classed as unique or problematic—including reactive materials, gas cylinders, and tritium-contaminated materials—comprise only about 10 percent of the inventory. They are often overlooked in site cleanup contracts, but they deserve special attention for research because some will be difficult to treat and may eventually become roadblocks to site closure. Application of biotechnologies for treating wastes has been largely overlooked.

The committee believes that the greatest challenges for the next generation of treatment technologies lie in developing

- emission-free treatment processes,
- treatments for problematic or unique wastes, and
- methods to ensure the long-term durability of stabilized waste.

Opportunities for basic research lie in chemical treatment, biological treatment, and waste stabilization. For chemical treatment, understanding the speciation of inorganic constituents, oxide-substrate interactions, and mechanisms of gas production and adsorption (especially hydrogen) is fundamental. In the biological area, research should include enzymatic or whole-cell approaches that target specific or broad categories of contaminants, biotransformations for removing mercury and heavy metals, bioaugmentation or biostimulation to remediate actinide-impacted soils, and development of hydrogen and methane scavengers. In the stabilization area, research should address new approaches to stabilizing buried waste prior to or in the early stages of excavation, smart materials that react with waste constituents, and very long term barriers against contaminant migration and methods to prove their longevity.

Public concern about air emissions from incineration has created incentives for applied research toward large-volume, robust alternatives that are emission free, as well as to smaller-scale, portable devices that may have specialized applications. There are also opportunities to develop more efficient processes that yield smaller or easier-to-manage waste streams from DOE's ongoing activities (e.g., isotope production, generation of secondary wastes from high-level waste processing, facility deactivation and decontamination).

Long-Term Monitoring

The EMSP should support research to improve long-term monitoring of stored and disposed TRU and mixed wastes. Research should emphasize remote methods that will help verify that the storage or disposal facility works as intended over the long term, provide data for improved waste isolation systems, and inform stewardship decisions.

To ensure safety, wastes and the facilities that house them have to be monitored. This includes monitoring during storage, which could continue for decades for some wastes, and during the operating life of the disposal facility. For example, substantial deformation of the salt will occur during the operational phase of WIPP, and monitoring can help DOE understand and verify how lithostatic forces will seal the disposal rooms. Very long term monitoring will continue after the disposal facility is closed.

DOE appears to have no firm plans for long-term monitoring of stored or disposed wastes. Research begun now can lead to reliable, cost-effective monitoring devices and methods. Data from monitoring can help ensure safety, reassure concerned citizens, and assist in the development of new disposal facilities.

The committee believes that the greatest challenges for the next generation of monitoring technologies lie in providing

- long-lived, reliable sensors (and power supplies) that can be remotely interrogated, and
- airborne or satellite imaging.

Research opportunities exist, for example, in developing smart sensors that self-analyze and report drum location and contents, and smart filters that monitor the type and amount of gas produced in a drum. In addition to being a repository, WIPP can be an important laboratory for repository science and sensor technology. Research should focus on potential biodegradation of the various organic components, reactions altering the geochemistry of the inorganic compounds, biogeochemical factors that affect leaching or migration of toxic and radioactive materials, and the effect of physical conditions and chemical composition on the biogeochemical processes occurring in the waste.

Concluding Comments

Accelerating site closure, a key feature of EM's planning since the 1990s, has been emphasized by EM's top-to-bottom review. DOE is presently making a concerted effort to remove TM wastes from its sites as rapidly as possible. Among the areas for EMSP research recommended by the committee, research in characterization that would expedite shipping wastes for off-site disposal is most likely to provide immediate payoffs. Research toward methods for treating wastes that do not meet shipping or disposal criteria might provide similar near-term payoffs.

Nevertheless, closing the larger DOE sites will require decades. Problems that are not foreseen or appreciated today are likely to be encountered in buried waste retrievals. Monitoring WIPP during its operational period is a unique scientific opportunity. Demonstrating that WIPP behaves as expected could be invaluable as DOE seeks to open other geological waste repositories. Buried waste retrieval and monitoring of disposal facilities provide opportunities for the long-term, breakthrough research envisioned by Congress, and these opportunities should not be overlooked in DOE's rush to meet short-term needs.

1

Introduction, Background, and Task

The Department of Energy's (DOE's) Environmental Management Science Program (EMSP) was established by the 104th Congress[1] to bring the nation's basic science infrastructure to bear on the massive environmental cleanup effort under way in the DOE complex. The objective of the EMSP is to develop and fund a targeted, long-term research program that will result in transformational or breakthrough approaches for solving the department's environmental problems. The goal (DOE, 2000a, pp. 1-2) is to support research that will

- Lead to significantly lower cleanup costs and reduced risks to workers, the public, and the environment over the long term.
- Bridge the gap between broad fundamental research that has wide-ranging applicability . . . and needs-driven applied technology.
- Serve as a stimulus for focusing the nation's science infrastructure on critical national environmental management problems.

To help meet these goals, the EMSP provides three-year competitive awards to investigators in industry, national laboratories, and universities to undertake research on problems relevant to DOE cleanup efforts. From its inception in 1996 through fiscal year 2001, the EMSP has provided $294 million in funding for 361 research projects.

This study, addressing transuranic and mixed wastes, is the fourth study undertaken by the National Research Council (NRC) to assist DOE in developing a research agenda for the EMSP.[2] The previous three reports gave advice for research in subsurface contamination, high-level waste, and facility deactivation and decontamination (NRC, 2000a,

[1]Public Law 104-46, 1995.

[2]An initial study advised DOE on establishing the EMSP (NRC, 1997a).

2001a, 2001b). DOE has used these studies in developing calls for research proposals and for evaluating submitted proposals. A fifth study, addressing excess nuclear materials and spent DOE nuclear fuel, is in progress (NRC, 2002a).

After its establishment by Congress and through most of the course of this study, the EMSP was managed through a partnership between the DOE Office of Environmental Management (EM), which has primary responsibility for the cleanup mission, and the DOE Office of Science, which manages basic research programs. The advice provided by the NRC studies, as well as the EMSP's calls for proposals, reflected EM's organization of its science and technology development activities into five "focus areas," which are the topical areas of the NRC studies mentioned above—subsurface contamination, high-level waste, facility deactivation and decommissioning, transuranic and mixed wastes, and nuclear materials (see also Appendix A).

As this report was being finalized, EM completed a top-to-bottom review of its organization, which was directed by the Secretary of Energy (DOE, 2002). As a result of the review, the Office of Biological and Environmental Research within the Office of Science will become solely responsible for administering the EMSP. The focus area structure under EM will be discontinued. Subject to approval by Congress, these changes will become final at the start of fiscal year 2003. As it finishes its work on this report, the committee[3] understands that the EMSP's previous approaches to issuing calls for research proposals, evaluating submitted proposals for both scientific merit and relevance to EM's needs, and funding the proposals will remain largely unchanged. Readers of this report who may intend to submit proposals to the EMSP should seek updated information from the DOE Office of Science.[4]

Statement of Task

The statement of task for this study charged the committee to provide recommendations for a science research program for managing mixed and transuranic wastes that are currently stored at DOE sites or will be produced as part of DOE's site cleanup program (see Sidebar 1.1).

To address the statement of task, the committee has made recommendations in four categories in which it believes that EMSP-funded

---

[3]The Committee on Long-Term Research Needs for Managing Transuranic and Mixed Wastes at Department of Energy Sites, which developed this report, is referred to as "the committee" throughout.

[4]See http://www.sc.doe.gov/production/ober/ober_top.html.

SIDEBAR 1.1 STATEMENT OF TASK

The objective of this study is to provide recommendations to the Department of Energy's Environmental Management Science Program for the development of a research agenda to address challenges in managing mixed and transuranic (TRU) wastes that are currently stored at DOE sites or will be produced as part of DOE's site cleanup program. The study will accomplish the following:

1. **Evaluate the next generation of treatment technologies and cleanup approaches for the specific categories of DOE TRU and mixed waste for which current treatment technologies are not adequate, in particular due to new or tightened regulatory requirements or other nontechnical considerations such as nascent public opposition to incineration.**
2. **Identify gaps in the scientific basis for selecting or implementing new treatment technologies.**
3. **Identify areas of research where EMSP can make significant contributions to solving DOE's mixed waste problems and add to scientific knowledge generally, taking into account research funded by other programs besides the EMSP.**

research is most likely to lead to significant new or breakthrough technologies: waste characterization, retrieval of buried wastes, waste treatment, and long-term monitoring. Characterizing wastes and treating them (as necessary) for shipment to disposal facilities are subjects of intense current efforts at DOE sites. However, the inventory of transuranic and mixed wastes is extensive, and work to dispose of this inventory will continue for 20 years or more, which provide time and incentive for significant research and technology development. Buried waste retrieval and long-term monitoring of waste disposal have received little attention within DOE, but they are likely to present significant obstacles for completing site cleanup.

Chapter 2 of this report frames DOE's broad challenges in managing and disposing of its transuranic and mixed wastes—the large and diverse inventory, multiple and changing regulations, and public concerns. Chapter 3 sets out the committee's research recommendations in each of the four categories described above.

The first subtask asks for an evaluation of next-generation treatment technologies in instances where current technologies may become inadequate for nontechnical reasons—an example being incineration, which was under review by a special DOE panel at the time this committee was chartered. The committee did not attempt to evaluate next-generation treatment technologies per se, but rather identified challenges (technical and nontechnical) likely to confront these next-generation technologies (see Chapter 3). The committee felt that this approach was more fruitful for providing guidance for an EMSP research agenda. Further, the committee concluded that any new technologies or changes to agreed-upon cleanup plans are likely to encounter public opposition unless the public is involved in the selection process (see Chapter 2).

In presenting its recommendations, the committee gives a brief discussion of current baseline technologies,[5] challenges for next-generation technologies (as discussed above), and research opportunities. Although the discussions were influenced to some degree by the backgrounds and expertise of committee members, the research recommendations were arrived at by a consensus process that considered input to the committee, site needs, the existence of critical knowledge gaps, the potential for future cost and schedule savings, and the possibility of achieving technology breakthroughs.

The committee held five meetings between May 2001 and February 2002 to gather information (see Appendix E). The committee's fact finding included site visits and briefings at the Idaho National Engineering and Environmental Laboratory, Oak Ridge Reservation (Tennessee), Savannah River Site (South Carolina), Hanford Site (Washington), and Waste Isolation Pilot Plant (New Mexico). The committee also received briefings by DOE headquarters personnel who administer the EMSP and by representatives of EM's Transuranic and Mixed Waste Focus Area.

---

[5]Baseline technologies are those that are being used at DOE sites or that are commercially available and included in DOE's site cleanup plans.

2

Framing DOE'S Transuranic and Mixed Waste Challenges

The accumulation of radioactive waste materials began in the 1940s with the development of the atomic bomb and continued with the large-scale refining and production of fissile materials such as uranium and plutonium during the Cold War. Processes included separation and enrichment of special isotopes, reactor fuel fabrication, dissolution and chemical separation of irradiated materials, and fabrication (casting, machining, plating) of weapons components. During this period, emphasis was placed on production and little attention was given to reducing the volume or variety of wastes. The wastes were managed using practices analogous to those used in other process industries, including on-site disposal in landfills and the use of ponds and lagoons to manage large volumes of wastewater.

Wastes generated by production operations ranged from slightly contaminated trash to highly radioactive liquids from processing irradiated fuels. Frequently these wastes contained both radioactive and hazardous chemical substances. This chapter provides a context for the Department of Energy's (DOE's) challenges in managing wastes contaminated with both hazardous chemicals and low levels of radioactive fission products (mixed low-level waste [MLLW]) and wastes contaminated with transuranic isotopes (TRU waste)—see Sidebar 2.1. Research challenges for managing DOE's high-level radioactive waste and spent nuclear fuels and for remediating subsurface contamination are described elsewhere (NRC, 2000a, 2001a, 2002a) and are not dealt with in this report.

During most of the time this study was in progress, the Transuranic and Mixed Waste Focus Area (TMFA), a part of the DOE Environmental Management Office of Science and Technology (EM-OST), was responsible for ensuring that technologies were available to manage this waste. Organizational changes within EM-OST that occurred as this report was being finalized are described in Appendix A.

SIDEBAR 2.1 WHAT ARE MIXED LOW-LEVEL AND TRANSURANIC WASTES?

The committee used the following working definitions in preparing this report. They are based on the EPA Mixed Waste Glossary (EPA, 2002a). As noted, they were derived from detailed definitions in Congressional acts or developed by the federal agencies that regulate these wastes: the DOE, the Nuclear Regulatory Commission (USNRC), and the Environmental Protection Agency (EPA).

Low-level radioactive waste (LLW) **is defined in the Low-Level Radioactive Waste Policy Amendments Act of 1985, essentially by excluding other types of waste. Namely, LLW is not spent nuclear fuel, high-level radioactive waste from reprocessing spent nuclear fuel, or byproduct material. Most wastes in the DOE inventory that are designated as LLW are contaminated with small amounts of radioactive fission products, which are the isotopes that result from splitting (fissioning) the uranium nucleus.**

Hazardous waste **is defined by the EPA in Title 40 of the Code of Federal Regulations, parts 260 and 261. This waste is toxic or otherwise hazardous because of its chemical properties. Waste can be designated as hazardous in any of three ways:**

- **It contains one or more of over 700 materials listed as hazardous by the EPA;**
- **It exhibits one or more hazardous characteristics, which include ignitability, corrosivity, chemical reactivity, or toxicity;**
- **It arises from treating waste already designated as hazardous.**

Mixed low-level waste (MLLW) **meets the above definitions of both low-level waste and hazardous waste. It contains materials that are chemically hazardous and low levels of radioactive contamination.**

Transuranic waste (TRU) **is defined by DOE Order 435.1 as waste that has a radioactivity of more than 100 nanocuries per gram that arises from alpha-emitting isotopes with atomic numbers greater than uranium (92) and half-lives greater than 20 years. Most TRU waste in the DOE inventory is contaminated with plutonium-239, which has a longer radioactive half-life (24,000 years) than most fission products.**

Mixed transuranic waste (MTRU) **meets the definitions of both transuranic and hazardous waste. EPA estimates that more than half of DOE's TRU inventory is MTRU (EPA, 2002a). Because all TRU wastes are destined for WIPP, DOE no longer distinguishes MTRU as a special category in its inventory (DOE, 2001a).**

The Department of Energy's challenges in managing and disposing its transuranic and mixed wastes (TM wastes) arise primarily from three factors. One is the large and diverse waste inventory, which is incompletely characterized. A previous study (NRC, 1999a, p. 18) of TM wastes found:

> *EM's mixed waste inventory is sufficiently characterized that conceptual design of treatment processes . . . can proceed. However, the inventory is insufficiently characterized for detailed engineering design of treatment processes or process optimization.*

Another challenge is the complex and evolving regulatory constraints that are applied to these wastes. The earlier study (NRC, 1999a, p. 22) noted:

> *The U.S. Environmental Protection Agency (EPA), U.S. Nuclear Regulatory Commission (USNRC), Department of Transportation (DOT), and individual states all exert measures of control over treatment, transport, and disposal of mixed waste. . . . [T]he range of regulatory approaches and resulting regulations create substantial challenges for treatment and disposal of mixed wastes.*

There is public concern about, and often opposition to, technologies that are unfamiliar or that might change agreed-upon cleanup plans. An international review of waste management programs (NRC, 2001c, p. 3) found the following:

> *Today the biggest challenges to waste disposition are societal. Difficulties in achieving public support have been seriously underestimated in the past, and opportunities to increase public involvement and to gain public trust have been missed.*

Based on its fact finding, the committee believes that these conclusions remain valid. Through their impact on site technology needs, challenges arising from the diverse waste inventory, multiple evolving regulations, and public concerns will significantly affect any research agenda developed by the Environmental Management Science Program (EMSP). These factors, which frame DOE's TM waste challenges, are discussed in this chapter.

DOE's Transuranic and Mixed Waste Inventory

Managing and disposing of DOE's TM waste inventory presents technical challenges and research opportunities because the inventory is large and diverse. This section provides an overview of the inventory with emphasis on wastes that led the committee to its research recommendations. Appendix B gives a detailed description of the inventory.

Inventory Description

Information on DOE's waste inventory is given in a summary report published in April 2001 (DOE, 2001a). DOE compiled much of the inventory data from its fiscal year 2000 Central Internet Database.[1]

---

[1]See http://cid.em.doe.gov.

FIGURE 2.1 Before 1970, transuranic and mixed wastes were buried in near-surface trenches. The waste was considered to be permanently disposed, and inventory data are lacking. Source: http://web.ead.anl.gov/techcon/images/ineel3.jpg.

TM wastes are described in two categories, transuranic and MLLW. The summary report does not distinguish between TRU and mixed transuranic waste (MTRU) (see Sidebar 2.1). All inventory data refer to the waste volume unless noted otherwise.

Since 1970, DOE sites have stored most TM wastes retrievably in 55-gallon drums or larger containers for future treatment, if needed, and disposal. Before 1970, DOE sites buried TM wastes in "shallow land" facilities, within about 30 meters of the surface.[2] Most waste was buried in 55-gallon drums, some was buried in other containers, and some had no durable container (e.g., burial in plastic bags, cardboard boxes, or without containment); see Figures 2.1 and 2.2. At the time, DOE generally considered buried waste to be permanently disposed. Recently, DOE has recognized that at least some of its buried waste inventory may require retrieval and treatment (DOE, 2001a).

Contaminated soils and sediments have resulted from previous DOE practices of discharging low-level liquid wastes to retention basins or

[2]A fraction was buried at "intermediate" depths between 30 and 300 meters.

FIGURE 2.2 Since 1970, DOE has required that sites store TRU waste so that it can be retrieved easily. TRU wastes at Hanford, which is in a very dry region, are stored in earthen mounds.
Source: DOE Richland Operations Office.

from leaks. DOE recognizes that some of these soils and sediments are sufficiently contaminated to warrant retrieval and describes these as "ex situ contaminated media" in its summary report. If they are retrieved, both the pre-1970 buried waste and the ex situ media will be considered newly generated waste (DOE, 2001a).

Table 2.1 gives an overview of DOE's current and expected inventories of TM wastes. Disposing of retrievably stored TRU waste, which contains an estimated 2.6 million curies of radioactivity, in the Waste Isolation Pilot Plant (WIPP) is a top priority for DOE (Triay, 2001). Buried TRU waste, with a volume comparable to the stored TRU, is estimated to contain about 400,000 curies. A large volume of buried MLLW is contaminated with alpha-emitting isotopes at levels below the regulatory threshold for TRU waste and is designated as α-LLW.[3] DOE expects to continue generating TRU waste until about 2034 and MLLW until about 2070, mainly from facility deactivation and decommissioning. In addition, DOE expects to produce ex situ waste by recovery of a portion of the more contaminated soils and sediments at some of its sites.

The diversity of the TM waste inventory is described in the Mixed Waste Inventory Report (MWIR [DOE, 1995]). This report was based on data compiled by DOE sites as a basis for developing their site treat-

[3]The radioactivity from alpha-emitting isotopes is estimated to be between 10 and 100 nanocuries per gram of waste.

TABLE 2.1 Overview of DOE's Transuranic and Mixed Wastes

	Volume	
Origin	TRU (m^3)	MLLW (m^3)
Buried (pre-1970)	137,000	317,000[a]
Retrievably stored (1970-1999)	111,000	44,500
Predicted new waste generation	60,000[b]	100,000[c]
Recovered soils and sediments (2002-2010)	32,000	170,000

[a] α-LLW.
[b] 2000-2034.
[c] 2000-2070.

SOURCE: DOE, 2001a.

ment plans as mandated under the Federal Facility Compliance Act of 1992. The inventory was divided into five treatment groups: debris, inorganic homogeneous solids and soils, organics, unique wastes, and wastewaters (see Sidebar 2.2). The treatment technologies for these groups were reviewed in a previous NRC (1999a) report.

Table 2.2 shows the relative amounts of retrievably stored wastes that fit into each of the treatment groups. Debris waste, which is very heterogeneous, comprises by far the largest category. Unique wastes make up a small fraction of the inventory. However, many unique wastes are problematic to treat and dispose, and their small volumes make them economically unattractive to site cleanup contractors.[4]

No information is available concerning the treatment needs for the previously buried waste. DOE's production processes did not change with the prohibition of burial in 1970, so these materials are expected to have a composition similar to retrievably stored waste. The distribution profile of wastes into the treatment groups is unlikely to change appreciably if buried wastes are retrieved.

The 1995 inventory also indicates DOE's level of confidence in how well the wastes were characterized. In general terms, DOE has high or medium confidence that the physical nature (i.e., soil or sludge) of most wastes is correctly identified but lacks confidence in the existing quantitative data on the wastes' chemical and radioactive constituents (see Appendix B for details).

[4]The TMFA recognized that unique wastes could become an obstacle to site closure and formed a Waste Elimination Team to identify and plan disposition of these orphan and hard-to-treat wastes (Hulet, 2002).

SIDEBAR 2.2 DIVERSITY OF TM WASTES

For the purpose of developing site treatment plans for TM wastes, DOE established five treatment groups. The types of waste included in each group provide a perspective on the overall waste diversity.

Debris

- ***Metal Debris:*** **Metal with or without lead or cadmium**
- ***Inorganic Nonmetal Debris:*** **Concrete, glass, ceramic or brick, rock, asbestos, and graphite**
- ***Organic Debris:*** **Plastic or rubber, leaded gloves or aprons, halogenated plastics, nonhalogenated plastics, wood, paper, and biological matter**
- ***Heterogeneous Debris:*** **Composite filters, asphalt, electronic equipment, and other inorganic and organic materials**

Inorganic Homogeneous Solids and Soils

- ***Inorganic Homogeneous Solids:*** **Particulate matter—such as ash, sandblasting media, inorganic particulate absorbents, absorbed organic liquids, ion-exchange media, metal chips or turnings, glass or ceramic materials, and activated carbon**
- ***Inorganic Sludges:*** **Wastewater treatment pond, off-gas treatment, plating waste, and low-level reprocessing sludges**
- ***Other Inorganic Waste:*** **Paint waste (chips, solids, and sludges), salt waste containing chlorides, sulfates, nitrates, metal oxides or hydroxides, and inorganic chemicals**
- ***Solidified Homogeneous Solids:*** **Soil, soil/debris, and rock/gravel**

Challenges in Managing the Inventory

The current and projected volume of TRU waste will pose significant challenges for disposing of this waste. Several hundred thousand drums will have to be shipped to WIPP (see Table 2.1). The characterization required for shipping and acceptance at WIPP currently requires several hours and costs about four thousand dollars for each drum (DOE, 2001d).[5]

[5]One cubic meter is equal to five 200-liter (55-gallon) drums, although WIPP can receive containers larger than 55-gallon drums.

Organics

- ***Organic Liquids:*** **Aqueous streams containing both halogenated and nonhalogenated organic compounds as well as pure organic streams containing halogenated and nonhalogenated compounds**
- ***Organic Homogeneous Solids:*** **Organic particulate matter (resins, organic absorbents), organic sludges (biological, halogenated, and nonhalogenated), and organic chemicals**

Unique Waste

- ***Lab Packs:*** **Organic, aqueous, and solid laboratory chemicals and scintillation cocktails**
- ***Special Wastes:*** **Elemental mercury, elemental hazardous metals (activated and nonactivated lead, elemental cadmium), beryllium dust, batteries (lead acid, mercury, cadmium), reactive metals (bulk and reactive metal-contaminated components), pyrophoric fines, explosives or propellants, and compressed gases and aerosols**
- ***All Others:*** **Materials placed in a final waste form are included in this category**

Wastewaters

- **Acidic, basic, and neutral aqueous liquids and slurries, including cyanide-containing wastewaters and slurries**

Source: DOE, 1995.

Methods to streamline characterization are likely to save large amounts of time and money (see Chapter 3).[6]

Characterizing and treating MLLW, which has received relatively little attention compared to TRU waste, to meet Resource Conservation and Recovery Act (RCRA) disposal requirements will be a challenge. In spite of the lack of quantitative chemical characterization, most of the

[6]Compositions of waste generated after about 1999 are well documented according to requirements of the WIPP permit (see next section). Additional characterization of this waste should not be necessary. TRU wastes will be generated until about 2035 (DOE, 2001a).

TABLE 2.2 Distribution (percent) of Inventoried Waste in Treatment Groups

Group	TRU	MTRU	MLLW
Debris	95	71	57
Solids and soils	2	28	25
Organics	2	0.5	4
Unique	1	0.1	4
Wastewaters		0.1	9

NOTE: The MWIR distinguishes MTRU from TRU waste. About 2-4% of TRU and MTRU waste require remote handling.

SOURCE: DOE, 1995.

TABLE 2.3 Difficult-to-Treat Hazardous Components in DOE MLLW

	Percent of the Treatment Group that is Contaminated				
Type of Contamination	Debris	Organic	Solids and Soils	Unique	Wastewater
Metals	70	79	90	66	98
Solvents or other organics	77	90	75	23	27
Mercury	20	34	31	17	70

SOURCE: DOE, 1995

MLLW inventory is known to contain chemicals that are difficult to treat—heavy metals, solvents and other organics, and mercury (see Table 2.3). Further, there is considerable comingling of these classes of waste materials, making the selection of treatment options complicated.

Some components in TRU waste are problematic for shipping or disposal in WIPP (see Appendix B). About half of DOE'S TRU waste contains organic materials that have posed shipping problems due to potential gas generation, especially hydrogen. However, recent revisions to the Safety Analysis for TRUPACT-II shipping containers have reduced but not eliminated the concern about hydrogen accumulation during shipment. Under the new revision, only about 2 percent of the TRU waste inventory (about 14,200 drum equivalents) continues to face shipping restrictions.[7] Reactive and corrosive chemicals (including paint

[7]Revision 19 to the Safety Analysis Report for Packaging for the TRUPACT-II (Curl et al., 2002).

FIGURE 2.3 Manual sorting of waste inside a containment (glovebox) facility is required to remove items that are prohibited by shipping or disposal restrictions. Sorting and repacking the waste are time-consuming, expensive, and present risks to workers.
Source: DOE Richland Operations Office.

spray cans, which are often found in waste drums) cannot be accepted at the WIPP, and they are removed by sorting through the waste (see Figure 2.3). Waste that is contaminated with polychlorinated biphenyls (PCBs), about 1 percent of the inventory, cannot currently be accepted by the WIPP.

Approximately 2 to 4 percent of the TRU waste inventory produces enough penetrating radiation from fission product contaminants that it requires remote handling (RH-TRU), rather than hands-on operator contact. The requirement for remote handling greatly increases the difficulty of characterizing, treating, and packaging or repackaging this waste. Meeting per-drum limits on heat generation and fissile material content can require repackaging the waste (Curl et al., 2002; Moody, 2002). In addition to increasing the waste volume, repackaging to meet drum limits is expensive, time consuming, and creates a potential for worker exposure.

Current and Evolving Regulatory Constraints

All waste handling and disposal operations are governed by regulatory requirements. However, DOE faces a particular challenge in managing TM waste due to the number of agencies that regulate this waste and the generally prescriptive nature of their regulations. At the federal level, TM wastes are the regulatory responsibility of DOE, the Environmental Protection Agency, and the U.S. Nuclear Regulatory Commission. Department of Transportation requirements apply to shipping the waste as well as packaging the waste for shipment.

The Federal Facility Compliance Act of 1992 (FFCA) requires that DOE facilities comply with all federal, state, and local laws and regulations pertaining to hazardous waste. TM waste is thus subject to hazardous waste requirements promulgated by EPA under the Resource Conservation and Recovery Act of 1976 and subsequent revisions. The EPA has delegated its authority to many states, which may add additional requirements of their own.

The FFCA did not alter the separation between DOE and the USNRC. DOE is legally self-regulating for radioactive wastes (or the radioactive components of wastes) according to the Atomic Energy Act of 1954. However, DOE follows USNRC guidelines as a practical matter.[8] Additionally, the USNRC has licensing authority over commercially operated waste disposal facilities in which DOE is disposing of MLLW. For some of this waste, the states regulate in place of the USNRC.[9]

Transuranic Waste

Currently, DOE's TRU waste disposal efforts are focused on maximizing the utility of the Waste Isolation Pilot Plant, which is located deep underground in a salt formation in southeastern New Mexico. In 1992, the WIPP Land Withdrawal Act transferred control of the land at the site from the Department of Interior to the DOE. Subsequent amendments exempted WIPP from RCRA treatment standards and land disposal regulations (NRC, 1996).

WIPP operates under a permit issued by the State of New Mexico, which allows it to receive only TRU waste resulting from the nation's defense programs. DOE has committed in its permit application to manage all TRU waste as though it were mixed waste. In fact, the WIPP Waste Acceptance Permit (the Permit) specifically prohibits DOE from

---

[8]DOE Order 435.1 Radioactive Waste Management meets and extends provisions of USNRC waste management and radiation protection regulations, which are described later in this section.

[9]Under the Low-Level Radioactive Waste Policy Amendments Act of 1985, an "agreement state" is a state that has entered into a formal agreement with the USNRC and has the authority to regulate disposal of low-level radioactive waste within the state.

disposing non-mixed TRU waste unless the waste has been characterized in compliance with applicable provisions of the Permit. This is to avoid any question of New Mexico's having authority to regulate radioactive waste that is not subject to RCRA.

The Permit recognizes two classes of TRU waste: retrievably stored and newly generated. Retrievably stored refers to waste generated after 1970 but before the characterization requirements of the Permit were implemented at DOE sites (in about 1999). Newly generated refers to waste generated more recently. If wastes buried before 1970 or contaminated soils are retrieved, they will be considered as newly generated waste upon retrieval (see Table 2.1). Within each waste class, the Permit further categorizes three broad groups related to the physical form of the waste: homogeneous solids, soils and gravels, and heterogeneous debris (see Table 2.2).

Under the Permit, every retrievably stored waste container undergoes either radiography or visual examination to identify the physical form of the waste and to ensure that prohibited materials are absent.[10] Headspace gas analysis to determine the presence of volatile organic compounds (VOCs) must be performed on every container. Containers are assayed to be sure that their heat generation and fissile material content are within Permit limitations. In addition, some homogeneous solids and soil or gravel wastes must be sampled to establish the concentrations of VOCs, semi-VOCs, and metals for hazardous waste characterization.

Currently, the Permit is limited to wastes that produce a radiation dose rate of less than 200 millirem per hour at the surface of the container. This waste is called contact-handled TRU waste (CH-TRU) because it is deemed safe for direct handling by workers. Waste that produces more then 200 millirem per hour, about 2 to 4 percent of the TRU inventory, is designated remote-handled TRU waste. Because RH-TRU presents a potential hazard to workers, DOE is seeking regulatory changes to simplify its characterization. The State of New Mexico and the EPA have not yet approved a DOE plan to characterize RH-TRU waste.[11] As noted later in this chapter, EMSP research will be especially important if DOE's expected regulatory changes to simplify characterizing RH-TRU and dealing with other problematic wastes are not forthcoming.

---

[10]Prohibited materials include liquids, compressed gases, PCBs in concentrations of 50 parts per million or more, and ignitable, corrosive, or reactive materials.

[11]Another NRC committee is assessing characterization requirements for remote-handled TRU waste (NRC, 2002b).

Mixed Low-Level Waste

Unlike TRU waste, MLLW has no special exemptions from regulatory controls. DOE is relying primarily on private contractors and commercial facilities to meet EPA and USNRC requirements for treating and disposing of its MLLW. MLLW cannot be disposed in WIPP because it does not qualify as TRU waste.[12]

The EPA has developed regulations for hazardous waste management and disposal principally under the authority of RCRA enacted in 1976.[13] RCRA has been amended several times, with the most significant amendments passed in 1984 as the Hazardous and Solid Waste Amendments. RCRA provides for cradle-to-grave control of hazardous wastes by imposing management requirements on generators and transporters of hazardous waste and on owners and operators of treatment, storage, and disposal facilities.

The Comprehensive Environmental Response, Compensation, and Liability Act (CERCLA, also known as Superfund) of 1980 addresses threats to public health and the environment from abandoned or active sites contaminated with hazardous or radioactive materials. Reauthorized by Congress in the Superfund Amendments and Reauthorization Act (SARA) of 1986, CERCLA gives EPA the authority to require remediation of hazardous waste previously disposed at DOE sites. Compliance with CERCLA may require the retrieval of some previously buried mixed wastes.

The EPA's hazardous waste regulations apply to more than 500,000 companies and individuals throughout the United States (Case, 1991). Thus, the EPA uses a prescriptive approach to develop regulations that are almost universally applicable and contain straightforward numerical criteria that are relatively easy to understand and enforce. The EPA defines hazardous waste, specifies treatment standards that must be met prior to disposal, and specifies standards for construction and operation of hazardous waste disposal sites. For DOE MLLW, which includes relatively small quantities of many wastes that are diverse and heterogeneous, this universal prescriptive approach poses problems.

A Memorandum of Understanding (MOU) between DOE and EPA to help resolve these problems was signed in February 2000 (Eaton and Carlson, 2002). Under the auspices of this memorandum, DOE and EPA have established several joint agency work groups to address issues

---

[12]The basis for excluding MLLW is legal rather than technical.

[13]A history of EPA regulation of mixed waste beginning in 1976 can be found on the EPA Mixed Waste Team home page: http://www.epa.gov/radiation/mixed-waste/.

such as alternatives to incineration, mercury-waste treatment and disposal, HEPA (high-efficiency particulate arresting) filter monitoring, and generally difficult technical issues where mixed waste does not fit well with the land disposal restriction treatment standards. The MOU also lists a number of recent EPA regulations that are likely to affect DOE's plans and technical needs for managing MLLW (see Sidebar 2.3).

USNRC regulations that affect the management of MLLW include the Low-Level Waste Disposal Regulations (10 CFR 61) and Radiation Protection Standards (10 CFR 20). The USNRC regulates the radioactive characteristics of low-level waste materials acceptable for near-surface land disposal through a combination of prescriptive and performance-based requirements. Performance assessment is required to calculate worker and public exposure risks associated with waste disposal. According to the USNRC, a near-surface disposal facility is one in which radioactive waste is disposed within the upper 30 meters of the land surface. Institutional control of access is required for 100 years, and within 500 years, wastes must decay to a sufficiently low level that the remaining radioactivity will not pose unacceptable hazards to an intruder or the general public.

To meet this latter requirement, further prescriptive regulations define three classes of waste that are deemed suitable for near-surface disposal. Classification as Class A (the easiest to dispose), Class B, or Class C depends on which radionuclides are present and their concentrations (see Table 2.4). If the waste qualifies as TRU or is contaminated above certain limits with long-lived radionuclides, it is not suitable for near-surface disposal.[14]

DOE expects to use Envirocare's Utah facility to dispose of about 30 percent of its MLLW (DOE, 1997). This is a commercial facility located in Tooele County, Utah, which is permitted for the disposal of several types of waste. This facility also provides some treatment capabilities, including stabilization by converting the waste to a solid material, macroencapsulation, and microencapsulation (see Chapter 3).

The State of Utah has permitting authority for low-level waste and hazardous waste using USNRC and EPA rules, respectively. Currently the facility is licensed to receive only USNRC Class A radioactive waste and naturally occurring or accelerator-produced material. Disposal of Class B or C waste requires additional approvals by the Utah Radiation Control Board (already issued), the governor, and the Utah legislature. However, it is not clear at this time if Envirocare will pursue these approvals.

---

[14]Mining industry waste is excluded from this requirement.

SIDEBAR 2.3 ENVIRONMENTAL PROTECTION AGENCY (EPA) ACTIVITIES AFFECTING DOE MIXED WASTE

A number of EPA's regulatory activities that are under way or have been completed recently will impact mixed waste storage, treatment, and disposal. Research and development are necessary to support the development of emerging rules and to comply in a cost-effective manner with rules that have been finalized. Examples of regulatory activities that likely will drive research and development needs within the next three to five years include the following:

Mercury Hazardous Waste Treatment Standards—Notice of Data Availability. **The EPA, working with DOE, is evaluating technologies to stabilize mercury-containing wastes that are not suited for mercury recovery and elemental mercury stocks. These studies will describe the conditions under which various treatment process residues may remain stable in a landfill over the long term. The data report is being prepared and will be subjected to peer review.** ***A Notice of Data Availability containing the data and the peer review results is expected in late 2002.***

Mercury Action Plan. **This consists of an assembly of potential regulatory and voluntary actions, enforcement and compliance, research, and outreach to characterize and reduce risks associated with mercury. Its multimedia and cross-discipline focus and its emphasis on pollution prevention will impact mixed wastes containing mercury.** ***Estimated completion is expected in late 2002.***

Hazardous Waste Combustion Emission Standards. **On September 30, 1999, EPA promulgated standards to control emissions of hazardous air pollutants from incinerators, cement kilns, and lightweight aggregate kilns that burn hazardous wastes (referred to as the Phase I rule). A number of parties, representing the interests of both industrial sources and the environmental community, sought judicial review of the rule. On July 24, 2001, the United States Court of Appeals for the District of Columbia Circuit granted the Sierra Club's petition for review and vacated the challenged portions of the rule. On October 19, 2001, after several months of negotiation, EPA, together with all other petitioners that challenged the hazardous waste combustor emission standards, filed a joint motion asking the court to stay the issuance of its mandate for four months to allow time to develop interim standards. These stopgap interim standards were promulgated on February 13 and 14, 2002. They replace the vacated standards temporarily, until revised replacement standards are promulgated in 2005 through a full**

Along with using commercial disposal facilities, DOE sites can establish on-site facilities. Both the DOE Hanford Site and the Nevada Test Site are developing RCRA-compliant facilities for their own wastes and might receive waste from other sites in the future (see Figure 2.4).[15]

In summary, MLLW that contains certain specified materials is pro-

---

[15]Hanford's MLLW facility is operating under an interim permit. Hanford expects to be fully permitted to accept MLLW in about 2003. The Nevada Test Site expects to have permits in about 2004 (Maio and Reese, 2002).

notice-and-comment rule making that complies with the court's mandate. Also, EPA is developing Maximum Achievable Control Technology (MACT) standards for hazardous waste burning boilers and hydrochloric acid production furnaces as a second phase (Phase II) of the hazardous waste combustor (HWC) National Emission Standards for Hazardous Air Pollutants (NESHAP). ***DOE facility compliance date for the interim standards is September 30, 2003.***

PCB "Mega-Rule". **On June 29, 1998, EPA promulgated amendments to the regulations in 40 CFR 761 that significantly affect the use, manufacture, processing, distribution in commerce, and disposal of PCBs. This Mega-Rule affects mixed wastes containing PCBs. Among other things, the amendments provide new alternatives for the cleanup and disposal of PCBs, establish standards and procedures for decontaminating materials contaminated with PCBs, and create a mechanism for recognizing, under the Toxic Substances Control Act, other Federal or State waste management permits or approvals for PCBs.** ***The rule became effective in August 1998.***

LDR Phase IV and Progeny. **On May 26, 1998, EPA promulgated treatment standards for characteristic metal-bearing wastes, including mixed wastes, under the RCRA Land Disposal Restrictions (LDR) program. The regulations also adopted alternative treatment standards for soil contaminated with hazardous waste. On May 11, 1999, this rule was corrected and clarified, particularly with respect to treatment residuals and point of generation—both of which directly affect DOE mixed waste facilities.** ***DOE facility compliance date was August 1998 for metal standards; authorized state programs control the effective date of soil treatment standards.***

Hazardous Waste Identification Rule. **On May 16, 2001, EPA published a final rule, known as the Hazardous Waste Identification Rule (HWIR) that retained, with revisions, the mixture rule and the "derived-from" rule in the RCRA regulations (66 FR 27266). The revisions to the mixture and derived-from rules exempt mixtures and/or derivatives of wastes listed solely for their ignitability, corrosivity, and/or reactivity characteristics and also conditionally exempt certain mixed waste from the mixture and derived-from rules.** ***Effective date of final rule was August 14, 2001.***

SOURCES: EPA Office of Solid Waste and DOE Office of Science and Technology.

hibited from near-surface disposal under current EPA and USNRC regulations. These include the following:

- liquids,
- reactive or explosive materials,
- flammable material,
- untreated biological material,
- materials that may emit toxic gases or fumes,
- other materials subject to EPA's LDR, as listed in 40 CFR 268,

TABLE 2.4 Allowable Concentrations of Short-Lived Radionuclides for Near-Surface Disposal

Radionuclide	Class A Waste (Ci/m^3)	Class B Waste (Ci/m^3)	Class C Waste (Ci/m^3)
Total of all nuclides with less than 5-year half-life	700	*a*	*a*
H-3	40	*a*	*a*
Co-60	700	*a*	*a*
Ni-63	3.5	70	700
Ni-63 in activated metal	35	700	700
Sr-90	0.04	150	7,000
Cs-137	1	44	4,600

a: There are no limits for these radionuclides in Class B or C wastes. Practical considerations such as the effects of external radiation and internal heat generation on transportation, handling, and disposal limit the concentrations for these wastes.

SOURCE: Code of Federal Regulations, Title 10, Part 61.55.

FIGURE 2.4 RCRA requirements for disposal of MLLW include use of an impermeable liner and leachate collection system to provide total containment of hazardous chemicals for at least 30 years. Here, a large box of macroencapsulated waste is being placed in a RCRA-compliant disposal facility at Hanford. Source: DOE Richland Operations Office.

and

- radioactive isotopes in amounts that exceed USNRC Class C.

In order to be disposed, these wastes require treatments that may be technically difficult and expensive, as described in Chapter 3.

Evolving Regulations

In establishing criteria for accepting waste at the Waste Isolation Pilot Plant, DOE attempted to combine complex regulatory programs with what were expected to be the performance characteristics of the WIPP system, even as WIPP was being designed and built. In hindsight, as actual operation experience is gained, some of the self-imposed and rather restrictive requirements are proving impractical and perhaps even irrelevant from a health and safety perspective, such as the PCB limitations and the lengthy characterization protocols. A previous NRC report (2001d) concluded that the requirements should be reviewed and updated so that the criteria (referred to as waste acceptance criteria, WAC) are kept relevant to long-term performance of the repository and to safety, technical, and legal considerations.

DOE has focused its efforts on simplifying the regulatory requirements for wastes that might be prohibited from disposal at WIPP or that might be sidelined due to failure to meet the WAC, particularly because of characterization difficulties. For example, DOE has prepared a draft request for authorization to allow the disposal at WIPP of TRU wastes containing PCBs. Approval of this request would allow DOE to dispose of approximately 88,000 cubic feet (2,500 cubic meters) of TRU wastes containing PCBs subject to regulation under the Toxic Substance Control Act (TSCA). In addition, DOE is also drafting requests for options for waste characterization being conducted at its sites that send waste to WIPP for management and disposal. DOE is particularly interested in simplifying the requirements for characterizing its RH-TRU waste (NRC, 2002b).

EPA requirements that will affect DOE's management of MLLW for the next three to five years are identified in the MOU (see Sidebar 2.3). Given EPA's broad responsibility to regulate hazardous waste, additional future regulations affecting MLLW are inevitable.

Public Concerns

The views and concerns of members of the public will play an important role in establishing needs for improved technologies or changing agreed-upon plans for managing TM wastes. The statement of task for this report included evaluating treatment technologies for categories of TM waste for which current treatment technologies are not adequate, "in particular due to new or tightened regulatory requirements or other non-technical considerations such as nascent public opposition to incineration" (see Sidebar 2.4 and Appendix C).

SIDEBAR 2.4 PUBLIC CONCERNS ABOUT INCINERATION FOR TREATMENT OF TM WASTE

DOE's recognition that the public might oppose some of its waste treatment technologies arose from a lawsuit over plans to construct an incinerator for TM wastes at the Idaho National Engineering and Environmental Laboratory. To settle the suit, DOE appointed a Blue Ribbon Panel of independent experts to identify technological alternatives to incineration that might become available for use at DOE facilities nationwide (DOE, 2000b).[1] Subsequently, DOE formed an Alternatives to Incineration Committee (ATIC) to follow-up the technical and public perception issues involving the proposed alternatives.[2] To assist the ATIC, the INEEL Citizens Advisory Board produced a list of some 44 concerns for ATIC to consider in evaluating alternatives to incineration. About half are listed below as examples of the range and detail of citizens' concerns.

- **Size (mobility) of facility**
- **Cost of facility**
- **Complexity of operation**
- **Temperature**
- **Pressure**
- **Hazardous reagents**
- **Energy efficiency**
- **Maturity of technology**
- **Availability (ability to implement) in the short term**
- **Air emissions**
- **Type(s) of waste generated**
- **Volume(s) of waste generated**
- **Validity of monitoring results**
- **Disposition of waste generated**
- **Effects on worker and public health and safety**
- **Environmental impacts**
- **Residual effects or impacts that cannot be mitigated**
- **Acceptability to Shoshone-Bannock Tribes**
- **Description of catastrophic failure or credible accident scenario**
- **Description of off-normal operation**
- **Vulnerability to off-normal operation**
- **Emissions resulting from off-normal operations**

---

[1] **Settlement Agreement: *Keep Yellowstone Nuclear Free v. Richardson, et al.;* No 99 CV1042J (D.WY).**
[2] **The co-chairman of ATIC, Victoria Tschinkel, is a member of the committee that developed this report.**

During its site visits the committee heard from citizen groups at the Idaho National Engineering and Environmental Laboratory (INEEL), the Oak Ridge Reservation, and the Savannah River Site (see Appendix E). The committee did not hear a consistent opposition to incineration, but it did hear concerns about air emissions, buried waste retrieval, and monitoring. These broader concerns indicated that citizens are generally well informed about potential technology-related problems near their communities. Importantly, they expected DOE to address their concerns.

The committee believes that the key is not simply to develop new technologies to replace those that have raised public concern—to try to stay technologically one step ahead of the public—but rather to involve the public in the selection of technologies. The need for public participation is well documented (Chopayk and Levesque, 2002; Cohn, 2002; see also Busenberg, 1999). An earlier NRC study emphasized the importance of involving the public in choosing among technical options (NRC, 2001c, p. 24):

> *The challenge is therefore not just to identify options that are deemed suitable by technical experts. . . . Support for any chosen technology will be difficult to achieve unless options for managing wastes can be presented, together with their consequences, and the public can participate in choosing among those options.*

The committee agrees with this assessment. Any new technologies or changes in accepted cleanup plans are likely to encounter public concerns or opposition unless convincing scientific evidence for their adoption can be presented and citizens are involved in decision making. Providing a scientific basis for decision making that can be understood in the public forum is as important a role for the EMSP as providing routes to new technology.

Summary: Meeting TM Waste Challenges

DOE's efforts are focused on removing TM wastes from its sites as rapidly as possible. Although the focus is on near-term accomplishments, in the broader perspective DOE's waste inventory is large and diverse and wastes will continue to be produced by both site cleanup and new activities. During the course of DOE's several-decades-long, multibillion-dollar cleanup program, there are certain to be many changes as technology and regulations evolve and citizens express their concerns through the political process. There are both time and opportunity for EMSP research to produce new technologies to significantly enhance safety and reduce costs and uncertainties in DOE's management and disposal of its TM wastes.[16]

TRU waste disposal in the Waste Isolation Pilot Plant will require handling and characterizing hundreds of thousands of drums as well as larger containers during the next 20 or more years. DOE is working with pertinent regulators to reduce or eliminate restrictions that may not be necessary for reducing risk and that interfere with waste shipments or disposal. DOE expects this regulatory relief to help accelerate site cleanup and closure.

Conversely, DOE has encountered legal, economic, and public concerns about incineration, a technology that was expected to treat a large fraction of the TM waste inventory. Some buried wastes and contaminated soils, which DOE previously considered disposed, may have to be retrieved and treated as new wastes. Monitoring WIPP and other waste disposal facilities will continue for many years.

In view of the changes and challenges that DOE will be facing for decades, there are two clear roles for EMSP-funded research in DOE's TM waste management efforts:

1. To provide the scientific basis for new technologies that will be necessary for improving management and disposal of TM wastes during at least the next 20 years, especially if the regulatory changes that DOE expects to simplify dealing with problematic wastes are not forthcoming.
2. To enhance the scientific information available for regulatory decision making and public involvement, including evidence that disposal systems are operating as intended.

---

[16]This committee follows previous committees in noting that adequate research funding is a prerequisite for realizing benefits of new technologies (NRC 2000a, 2001a, 2001b, 2001e).

3

Research Needs and Opportunities

In this chapter the committee offers its views and recommendations on research opportunities for the Environmental Management Science Program (EMSP) to address challenges in managing transuranic and mixed low-level wastes. Based on its discussion of the issues that frame these challenges in Chapter 2 and its visits to Department of Energy (DOE) sites (Appendix E), the committee concluded that the most significant needs and opportunities lie in

- waste characterization and how waste characteristics may change with time,
- location and retrieval of buried wastes,
- waste treatment, and
- long-term monitoring.

The committee has been selective in its recommendations to encourage the EMSP to concentrate its limited funding in a few specific areas where the committee believes research can lead to the most significant improvements. Some technology areas, although clearly important, were excluded because in the committee's view the science and technology base already exists to address them on a relatively short time scale—less than five years.

Each recommendation is illustrated with a brief discussion of the current baseline technologies and technology gaps,[1] challenges for next-generation technologies, and research opportunities. Some examples are included, but these should not be construed as the only opportunities that the research community might perceive. Although the selection of examples was influenced to some degree by the back-

[1]Baseline technologies are those that are being used at DOE sites or that are commercially available and included in DOE's site cleanup plans.

grounds and expertise of committee members, the research recommendations were arrived at by a consensus process that considered input to the committee, the needs of the end user,[2] the existence of critical knowledge gaps, the potential for future cost and schedule savings, and the possibility of achieving technology breakthroughs.

Characterization

The EMSP should support research to improve the efficiency of characterizing DOE's TRU and mixed waste inventory. This should include research toward developing faster and more sensitive characterization and analysis tools to reduce costs and accelerate throughput. It should also include research to develop a fuller understanding of how waste characteristics may change with time (chemical, biological, radiological, and physical processes) to aid in decision making about disposition paths and to simplify the demonstration of regulatory compliance.

Waste characterization is defined as "the determination of the physical, chemical and radiological properties of the waste to establish the need for further adjustment, treatment, conditioning, or its suitability for further handling, processing, storage or disposal" (IAEA, 1993, p. 52). Current regulations require detailed characterization of waste for shipping and disposal, especially for transuranic (TRU) wastes destined for the Waste Isolation Pilot Plant (WIPP) as described in Sidebar 3.1. Characterization is also necessary to determine treatment options (see "Treatment" section later in this chapter). Information needed for waste characterization generally includes the identity and amount of radionuclides, liquids, volatile organic compounds (VOCs), polychlorinated biphenyls (PCBs), mercury, other metals regulated by the Environmental Protection Agency (EPA), and the rate of hydrogen generation. Such characterization increases the time and cost of preparing waste for off-site shipment as well as the potential for worker exposure to radiation.

Most transuranic and mixed wastes (TM wastes) are packaged in 55-gallon drums—hundreds of thousands of them as noted in Chapter 2. In addition, the Transuranic and Mixed Waste Focus Area (TMFA)[3] esti-

---

[2]End users are those who will use a given method or technology to accomplish a task. They are usually contractor personnel at DOE sites.

[3]During most of the time this study was in progress, the TMFA, a part of the DOE Environmental Management Office of Science and Technology (EM-OST), was responsible for ensuring that technologies were available to manage this waste. Organizational changes in EM that occurred as this report was being finalized are described in Appendix A.

mates that there are at least 12,500 large containers at five DOE sites that present special challenges for characterization because of their size. The sites cannot be closed without dispositioning these containers.[4]

One type of characterization is physical—an assembly-line style of nondestructive examination (NDE) and assay (NDA) to identify the contents of an unopened drum (see Figure 3.1). Is it sludge or debris, plastic or metal, solid or liquid? Has it been sealed, stabilized or treated properly? NDE methods also permit assessment of the heterogeneity of the drum contents and provide the means to screen for prohibited items (e.g., gas cylinders). A second type of characterization determines the chemical and radiological composition of the waste. Does the composition of the waste meet transportation and regulatory requirements for disposal? If not, how should the waste be treated? While these types of characterization provide snapshots of the waste, another consideration is that waste characteristics will change with time through radiological, chemical, and biological processes. The potential for gas generation, particularly hydrogen, is of concern. Understanding how the characteristics of containerized waste may change with time is especially important for its continued storage, shipping, and eventual disposal.

Baseline Technologies and Technology Gaps

For heterogeneous debris (see Chapter 2), the baseline characterization methodology for contact-handled TRU (CH-TRU) wastes comprises a number of steps, including radiography and opening the container for visual inspection to validate process knowledge (see Sidebar 3.1). Swipes or sampling and analysis are required to obtain contaminant information. If NDA is required, waste must be repackaged into containers sized for the available instrumentation. For homogeneous solids, a statistical number of drums require coring and analysis (St. Michel and Lott, 2002).

For TRU wastes that must be handled remotely because they also contain substantial amounts of gamma-emitting isotopes (RH-TRU), the current baseline requires the same characterization steps as CH-TRU. DOE is seeking to change this requirement because of the difficulty of making such detailed characterization in remotely operated facilities and the increased risks of worker exposure (NRC, 2002b). From a cost-saving standpoint, DOE would like to characterize RH-TRU based on process knowledge only. More realistically, however, DOE believes that

---

[4]Several sites also have buried wastes or contaminated media. These materials may not be containerized or their containers may be breached. They are discussed in the section "Waste Retrieval."

SIDEBAR 3.1 BASELINE CHARACTERIZATION STEPS FOR TRU WASTE

The characterization steps described here were developed for contact-handled TRU waste and have been applied to TRU mixed waste. The methods, equipment, procedures, determination of uncertainty, and other protocols used at DOE sites to perform these characterizations were approved by the DOE Carlsbad Field Office, New Mexico Environment Department, and EPA. The major steps depicted are as follows:

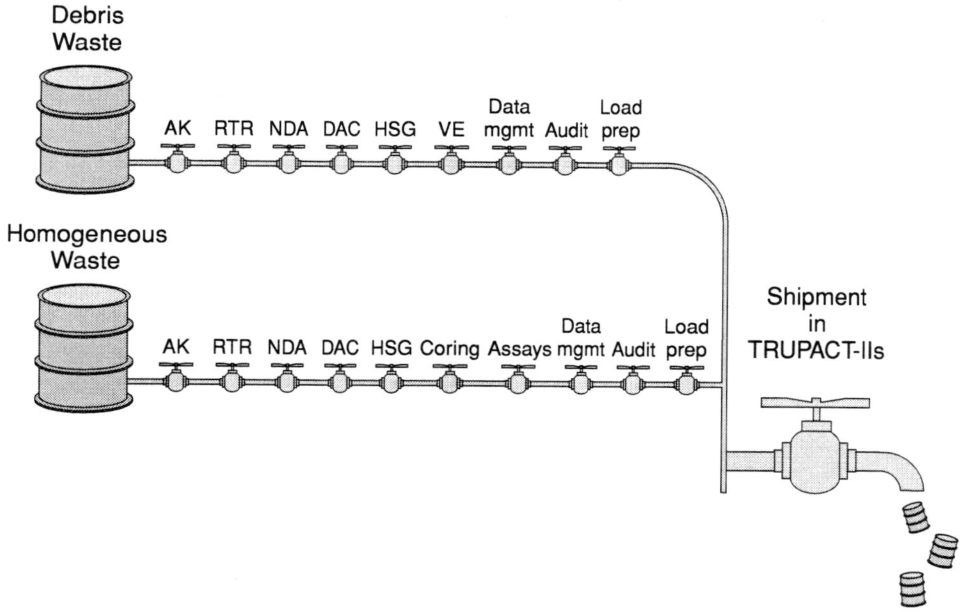

Determination of the Origin and Composition of the Waste by Acceptable Knowledge (AK). **Acceptable knowledge of the origin and composition of the waste must be documented to provide evidence that the waste has a defense origin (by the terms of the Land Withdrawal Act, only defense-related TRU waste may legally be sent to WIPP) and to provide characterization information on the waste constituents. The DOE Carlsbad Area Office and the EPA use acceptable knowledge documentation to certify each waste stream (i.e., waste-generating process). TRU waste sent to WIPP must come from a certified waste stream.**

Real-Time Radiography (RTR). **Radiography using X-rays is performed on all waste containers to look for items such as pressurized cans or free-standing liquids that are prohibited from being transported**

under U.S. Department of Transportation (DOT) regulations. If any of these items are present in a waste container, the prohibited materials are removed and the contents repackaged. This radiographic examination is also used to confirm the acceptable knowledge characterization information.

Radioassay and Determination of Fissile Isotope Content. The number of curies of each transuranic isotope is determined by radioassay (e.g., gamma scans) to a specified precision and accuracy. The fissile isotope content is assessed using nondestructive assay (NDA) methods, such as passive-active neutron systems. This information is used to meet the U.S. Nuclear Regulatory Commission (USNRC) requirement restricting the amount (several hundred grams) per container of each fissile species to ensure criticality safety.

Headspace Gas (HSG). Headspace gas sampling is used to check all waste containers for flammable gases (specifically, volatile organic compounds, hydrogen, and methane). This procedure, including resealing drums that have been vented and waiting specified times until gases regain equilibrium (drum aging to equilibrium criteria [DAC]), has been proposed as a means of ensuring conformity with the DOT (e.g., 40 CFR 173 and 40 CFR 177) and USNRC (e.g., 10 CFR 71) regulations that address the transport of flammable and/or gas-generating substances with radioactive materials. DOE proposed the HSG sampling procedure in its application to the USNRC for a licensing certificate on the transportation package (named the Transuranic Package Transporter, or TRUPACT-II) that is loaded with waste containers for transport by truck to WIPP.

Visual Examination (VE). A visual examination is performed on a fraction of the waste containers by placing the waste contents into a glovebox to verify the AK and RTR information. DOE proposed that 2 percent of the initial population of containers of each waste stream be examined visually, and if these evaluations resulted in few miscertifications, then the percentage of subsequent waste containers to undergo visual examination would be reduced. In October 1999, New Mexico in its Resource Conservation and Recovery Act (RCRA) Permit stipulated the initial fraction of containers to undergo visual examination to be 11 percent.

Coring and Assay of Homogeneous Waste for RCRA Constituents. Most of the TRU waste is heterogeneous in nature and requires no further characterization beyond acceptable knowledge to satisfy the regulatory requirements of RCRA. For homogeneous waste, a fraction of the waste containers (e.g., 55-gallon drums or standard waste boxes) are cored to extract representative samples that are analyzed for constituents (e.g., volatile and semivolatile organic compounds, toxic metals, other hazardous chemicals) regulated by RCRA.

SOURCES: NRC, 2002b, and DOE.

FIGURE 3.1 X-ray examination of a waste drum allows operators to determine if it contains prohibited items that must be removed before shipping the waste to a disposal site. Such visual inspection of a drum, at various positions and angles, may take several hours. Source: DOE Carlsbad Field Office.

some additional characterization will be required to validate the process knowledge (St. Michel and Lott, 2002).

DOE would like to simplify the characterization baseline for TRU wastes in order to increase the rate of shipping these wastes to WIPP. The main approach is to seek changes in current transportation and disposal requirements, for example, to reduce the many detailed characterization steps illustrated in Sidebar 3.1. In addition, the TMFA was developing improved characterization technologies. An example of state-of-the-art TMFA technology is the assay system being developed at the Idaho National Engineering and Environmental Laboratory (INEEL [see Sidebar 3.2]).

The committee believes that a gap exists in the lack of technologies available to automate sampling and characterization in a more reliable fashion. The problem becomes particularly complex for certain classes of wastes for which a single sample may not be representative (e.g., debris waste). Other technology needs include methods to nondestructively assay for radiological and nonradiological constituents (e.g., Resource Conservation and Recovery Act [RCRA] metals, low levels of TRU isotopes) and improved statistical methods to support approaches such as compositing.

SIDEBAR 3.2 STATE-OF-THE-ART TECHNOLOGIES BEING DEVELOPED BY THE TMFA

The Prompt Gamma Coincidence technique is a relatively new, nondestructive approach to measure isotopic ratios of plutonium and uranium. The technique overcomes the limitation of other approaches by measuring radiation associated with the fission process. When elements fission, a pair of fragments is produced. These fragments contain the same number of protons as the original isotope, emit neutrons and gamma rays, are in an excited state, and are short-lived.

The system uses coincidence measurement of gamma rays from the fission fragments. These gamma rays are distinct and are used to identify the fragment elements. Once the fragment elements are identified, the original, fissioning isotope can be confirmed. The system is a breakthrough technology because of its ability to distinguish among isotopes. This is important to determine the presence of weapon components and also to nondestructively qualify transuranic wastes and materials.

SOURCE: St. Michel and Lott, 2002.

Chemical and physical assays provide only a snapshot of the waste. An improved understanding of reactions that can change the waste characteristics with time is fundamental to making informed decisions on storing, shipping, and disposal. The Strategic Laboratory Council's analysis of DOE's environmental quality research and development portfolio found that the primary gaps in research in TRU and mixed waste disposal (DOE, 2000c, p. 28)

> *involve research to reduce uncertainties in waste and system performance driving conservatism in characterization and transportation requirements for TRU wastes. Improved performance knowledge may support reduction of characterization requirements, modified backfill requirements, and expansion of acceptable waste categories for WIPP.*

The current state of the art is that simple drum-by-drum methods are employed to empirically derive the thermodynamic and kinetic parameters for changes that occur in the waste as a function of broad waste categories. These methods often form the basis for assessment of the compliance of wastes with the waste acceptance criteria (e.g., gas generation rates, stability) for the pertinent disposal site.

There is a gap in basic knowledge of waste behavior, both in mechanistic understanding and in understanding other types of chemical behavior that may impact waste composition or physical integrity. These include the chemical form (speciation) of contaminants, metal-catalyzed redox reactions of waste constituents, sorption to the waste matrix, and effects of pH and ionic strength on the waste matrix. One example of unanticipated chemical reactivity is the recent demonstration that hydrogen can be generated by reduction of water by plutonium dioxide in oxygen-lean environments (Haschke et al., 2000). This serves as a pathway for hydrogen generation in addition to radiolysis, and has potential implications for how the plutonium might migrate from the waste.

There are gaps in understanding microbial effects in waste. Little has been done regarding the microbiology of TM wastes. Conversely, much has been done toward characterizing and exploiting the microbiota of more traditional hazardous wastes, resulting in significant cost savings over traditional disposal and remediation technologies (Harkness, 2000; Steffan et al., 2000). Microbial effects may be important in organic materials stored for long periods or in mixed waste landfills. Knowledge gaps include (1) what microbes are present in mixed and TRU waste and at what abundance; (2) what the activities of the microbes are and how their activity affects the waste material; and (3) how these microbes can be exploited to improve the treatment or disposal of TM wastes (Brockman, 1995; Newman and Banfield, 2002; Reysenbach and Shock, 2002).

In recent years, significant strides have been made toward developing methods and tools for analyzing microorganisms in environmental samples. Application of these techniques to TM wastes should lead to a better understanding of biogeochemical reactions occurring in the waste materials. This information will be important for assessing treatment options or monitoring the progress of biological treatment technologies applied to the material (Brockman, 1995; Newman and Banfield, 2002). Many of these methods and tools should be directly applicable for studying and characterizing the microbiology of TM wastes. For example, some of the more popular modern molecular biology-based techniques currently being used for environmental analysis include density gradient gel electrophoresis (DGGE [Muyzer, 1999]), terminal restriction fragment length polymorphism analysis (T-RFLP [Takai et al., 2001]), and whole or partial genome sequencing, but other equally useful methods clearly exist. Other studies have demonstrated that microbial DNA can be extracted from complex environmental samples, such as soil, and cloned to assess the genetic and functional diversity of uncultured organisms (Rondon et al., 2000). Still other technologies rely on identifying chemical signatures to confirm the presence of microorganisms or evaluate their activities in environmental samples (Nichols and McMeekin, 2002; Rütter et al., 2002; Zhang, 2002). Likewise, measuring the presence and abundance of specific biomarkers (e.g., peptides) can provide an understanding of microbial activities occurring in complex matrices (Elias et al., 1999). The combination of traditional culturing methods and modern molecular or chemical analytical techniques provides powerful tools for evaluating microbial populations in complex environments.

Challenges for Next-Generation Characterization Technologies

From its fact-finding visits to DOE sites and committee members' expertise, the committee believes that the greatest challenges for the next generation of waste characterization technologies will be to provide the following:

- more rapid, automated, NDA and NDE methods;
- more sensitive NDA and NDE technologies for larger containers and hard-to-detect chemical and radioactive contaminants; and
- improved methods, based on fundamental modeling, to derive present and future waste characteristics from a limited number of sampling parameters.

As uranium and transuranic elements such as plutonium decay, alpha particles are emitted. Associated with these alpha particles are mostly low-energy, nonpenetrating gamma rays and neutrons. Instruments for safeguarding nuclear materials rely on detecting these types of radiation. However, these measurements have drawbacks for waste assays because the radiation is subject to self-attenuation and shielding by extraneous materials, especially in larger containers. The resulting energy spectra are degraded to the extent that the resolution of currently available detectors is often insufficient to identify the specific fissioning isotope.

Multiple high-purity germanium detectors can be used to improve the sensitivity of gamma-ray spectroscopy (as can coincidence counting, see Sidebar 3.2), but this dramatically increases instrument expense and limits the number of stations that can be placed in service. Information about the spatial distribution of radioactivity within the containers is absent, which does not allow the detection of radiological "hot spots." Calibrating the results of such measurements may become complex and require statistical methods with assumptions about waste homogeneity and source location, which can introduce substantial error.

Neutron activation analysis identifies the chemical elements and thus can be used to search for transuranics noninvasively. It is, in fact, capable of identifying several radionuclides present that do not emit high-enough-energy photons to be readily detected using gamma-ray spectroscopy. Neutron activation analysis, however, suffers from the same turnaround, expense, calibration, and inhomogeneity problems associated with gamma-ray spectroscopy. Neutron activation analysis requires a high flux of neutrons, which is expensive, difficult, and potentially hazardous to provide. Inexpensive, bulk detection of TRU radionuclides by NDA presents a challenge for next-generation technologies.

DOE is presently developing new technologies for facility deactivation and decommissioning and subsurface contamination applications that are equally relevant for the characterization of containerized waste. Examples include portable "laboratory-on-a-chip" sensor technology for the quantitative identification of radionuclides and metals such as uranium, plutonium, cesium, strontium, mercury, and lead (Collins and Lin, 2001; Collins et al., 2002); the microcantilever sensor array technology for real-time characterization of the chemical, physical, and radiological content of ground water and mixed waste (Ji et al., 2000, 2001); and the micro-chemical sensor for in situ monitoring and characterization of volatile contaminants (Ho et al., 2001).

Similarly, the Department of Defense (DOD) has invested heavily in the development and deployment of both contact and standoff (non-contact) detectors for chemical and biological warfare agents.[5] The ultimate program goals are to develop robust, portable, real-time sensors capable of detecting agents well below incapacitating levels. The sensors are usually connected to air sample collecting or concentrating devices, although methods to analyze water and soil samples have also been developed. Chemical sensor technologies under development are based on ion mobility, surface acoustic wave (SAW), and miniature mass spectrometry. For biomolecules and organisms, sensor development includes fiber-optic waveguide, polymerase chain reaction (PCR), and DNA chip technologies. The present sensors are either chemical or biological only, but plans are to develop combined nuclear-biological-chemical sensors.

Microorganisms can potentially affect the chemical composition and physical properties of both the contaminants and the waste matrix. Radiation-resistant bacteria were first discovered in the mid-1950s (Anderson et al., 1956). The best-studied of these organisms is *Deinococcus radiodurans,* which can withstand up to 5,000 grays of gamma radiation without significant loss of viability (Battista, 1997). The bacterium has been used as a host organism to create radiation-resistant recombinant organisms for degrading pollutants (Lange et al., 1998) and for treating wastes contaminated with heavy metals (Brim et al., 2000). Its entire genome sequence has been elucidated (White et al., 1999). Recent studies suggest that many TM waste materials may

---

[5]For examples, see the articles in the special issue of *Biosensors & Bioelectronics,* Vol. 14 (2000).

contain viable and active microbial populations (Heitkamp, 2001). A previous National Research Council (NRC, 2001d) report summarized studies indicating that gas generation due to microbial degradation of cellulosic waste within WIPP will be insignificant, but it recommended monitoring.

The types of microbes present in TM wastes and the effect of microbes and microbial activity on the waste materials have largely been unstudied. These microbes could have profound effects on the ultimate fate of waste materials. The microorganisms may play a positive role by reducing the concentration of toxic organic constituents in the waste. They may play deleterious roles by increasing hydrogen or methane production, enhancing corrosion, or converting waste components into more toxic or mobile forms. Controlling or developing microorganisms to play a positive role is a challenge for future biotechnologies.

Research Opportunities

Research opportunities include noninvasive standoff imaging and image recognition methods and in-drum sensors to provide faster and more sensitive technologies for waste characterization. Research to develop predictive models of how waste characteristics may change with time, including microbial effects, can reduce the need for detailed waste analysis and provide better decision-making tools for storing, shipping, and disposing of TM wastes.

New Characterization Methods

Research toward new, nondestructive and noninvasive, characterization methods may lead to significant savings in time and cost, and decreased risk of worker exposure. While these methods are currently employed in the form of real-time radiography for physical characterization and various gamma and neutron imaging techniques for radionuclide inventory determination, basic research could also yield significant improvements in these methods, as well as means of identifying other constituents of waste drums. For example, it may be feasible to identify and image RCRA metal contaminants by devising new techniques in which neutron irradiation of the waste would activate stable metals, yielding products that could subsequently be mapped for concentration and location. Other approaches might include using the radioisotopes present in the waste drums to image the contents of the drum.

New methods might also include the use of alternative forms of energy to image constituents, ranging from the use of ultrasound to identify physical forms of waste to the use of customized imaging or local probes akin to magnetic resonance imaging (MRI) for the identification of a broad range of spin-active nuclei. Such methods, which employ detectors that are not sensitive to ionizing radiation, could avoid interference or background problems with waste that contains a significant amount of gamma-emitting isotopes, for example, RH-TRU. Promise exists for unique combinations of these methods with ionizing radiation detection methods using emissions (emitted radiation from the waste itself), induced emissions (radiation caused by external activation of waste contents), or transmissions (modification of energetic beams as they pass through waste). Similar measurement problems have been solved for medical and industrial applications. However, significant research is needed for these to meet the specific demands of the waste problem (e.g., spatial resolution, object sizes, heterogeneity, density, composition, field deployment).

Although noninvasive diagnostics are ideal, research also could improve the use of minimally invasive methods. Waste drums generally must be vented before they are shipped. This could provide a chance to emplace a variety of point detection sensors that would nondestructively convey information regarding waste constituents. When compared to conventional analytical methods that require withdrawing a sample, in situ probes could improve the speed of data acquisition and reduce associated secondary waste streams from the laboratory analyses. Examples of such probes could include fiber-optic windows for optical or spectroscopic characterization of drum contents ("optrode" approaches have been developed at Lawrence Livermore National Laboratory). Alternatively, one could envision the development of inexpensive chemical sensors operating on a variety of principles. The laboratory-on-a-chip and microcantilever sensors are examples of the type of sensor that could potentially be used for detecting changes in containerized and noncontainerized waste. EMSP projects could be coordinated with related DOD activities, especially in light of recent homeland security initiatives.[6] Additional research is needed to develop radionuclide spectrometers suitable for use as microdetectors in combination with miniaturized chemical characterization systems.

Automation and data handling could also speed the acquisition of analytical data. Current radiography techniques rely on time-consuming visual inspection by human operators to identify prohibited items in waste drums. With increasing sophistication of image recognition algorithms, further research could yield automated systems to improve the efficiency of real-time radiography (RTR) operations. Research opportu-

---

[6]See https://www.whitehouse.gov/homeland/.

nities exist in image interpretation, including self-attenuation and iterative reconstruction methodologies.

The success in application of analytical tools configured for high-throughput screening and analysis of combinatorial approaches to drug and materials discovery suggests that similar parallel approaches may be able to screen an array of large drums.[7] Wireless technology could also be used to track and monitor the drums remotely (see "Long-Term Monitoring" section). However, remote methods to introduce samples and provide long-term power have not been developed for these sensors. It may be possible to harvest electrical power to run these sensors directly from the thermal, chemical, biological, or radiological processes associated with the waste itself.

Microbial Effects

Research is needed to evaluate the microbiology of TM wastes. The research should focus on identifying the microorganisms that exist in the waste and evaluating their function relative to the waste material. The research should determine whether these microbes affect the hazardous or radioactive components of the waste in ways that make it more or less toxic or more or less suitable for disposal in hazardous waste, low-level waste, or other landfills or repositories (e.g., WIPP). Research could focus on specific processes such as gas (e.g., H_2, CH_4, CO_2) generation and utilization, corrosion, leaching, and biological and chemical transformation of hazardous and radioactive waste components. The research should evaluate the overall effect of physical conditions (e.g., pH, temperature, radioactivity) and chemical composition (e.g., organic and inorganic components, oxygen or other electron acceptor availability) on biogeochemical processes occurring in the containers.

Research also should evaluate the effects of microbial activity on waste forms for disposal, including polymers and grouts, macroencapsulation matrices, and containers. Additional research is needed to develop new tools for rapidly diagnosing microbial activity or identifying specific microbes in TM wastes. Evaluation of chemical signatures, biochemical markers, nucleic acid sequences, or other diagnostic characteristics can lead to sensor or detector technologies for rapid and even real-time characterization or monitoring. Such basic and applied research also might lead to new technologies suitable for use in many

---

[7]See "Combinatorial Discovery of Drugs and Materials," *Chemical & Engineering News,* March 8, 1999, p. 33.

areas including remote sensing, long-term monitoring, and even clinical diagnostics applications. The research should complement related research efforts being conducted within DOE and at other agencies (e.g., National Aeronautics and Space Administration's Astrobiology Institute research[8]).

Predictive Modeling

One of the most beneficial cost-saving tools in the management of TM wastes would be the formulation of more reliable predictive models of how waste characteristics may change with time, well validated by experimental data. Ideally, models could predict such factors as

- gas generation rates (e.g., matrix effects on rates of radiolysis, microbial effects);
- leachability of radioactive and hazardous constituents (are methods such as the toxicity characteristic leaching procedure [TCLP] accurate predictors of susceptibility to leaching?);[9] and
- the chemical availability of contaminants such as mercury for removal by various separation processes.

This information could simplify flowsheets, reduce the need for expensive drum-by-drum characterization, and improve the efficiency of waste packaging.

In order to construct a realistic model of hydrogen generation, for example, fundamental data would have to be compiled on rates of radiolysis for applicable organic constituents and water, rates of organic substrate diffusion under realistic conditions (clarifying matrix effects on rates of hydrogen generation), hydrogen diffusion and entrainment potential, and competing rates of chemical (e.g., recombination of hydrogen and oxygen catalyzed by metal or metal oxides) or biological reactions. These are undoubtedly complex models to derive and validate and will require new methodologies, including the means to couple interrelated parameters such as hydrogen availability (a complex function of generation and consumption, as well as physical diffusion) and metal ion oxidation states, which affect chemical reactivity.

There is a wealth of data from the actual sampling of each drum already sent to WIPP or ready to ship. A thorough analysis of these data might yield statistically valuable predictive tools. These tools might

---

[8]See http://astrobiology.arc.nasa.gov/.

[9]The TCLP is an EPA-prescribed test to determine whether a solid material should be classified as hazardous (see Chapter 2).

enable relatively inexpensive sampling of a few parameters in each drum to predict the presence of problematic materials. Alternatively, within a group of drums from the same source, sampling a certain percentage of the population may adequately predict the contents of the remainder.

New experimental approaches to validate these models may also be required. If such models could be employed in justifying higher wattage or organic content limits, it could result in tremendous cost savings (avoiding additional treatments to reduce organic content) and reduced risk of worker exposure (during repackaging).

Retrieval of Buried Waste

The EMSP should support research that will facilitate management of buried TRU and mixed waste in anticipation that retrieval of some waste will become necessary. This research should emphasize remote imaging and sensing technologies to locate and identify buried waste and retrieval methods that enhance worker safety.

Substantial quantities of TRU waste were disposed in near-surface excavations (shallow land disposal) prior to federal prohibition of TRU burial in 1970. Land disposal of untreated chemically contaminated wastes was not prohibited until enactment of the Resource Conservation and Recovery Act (RCRA) in the mid-1970s. Some of these wastes were buried in containers that may be retrievable;[10] some were buried in bulk. In addition, a quantity of pond and lagoon sludges and associated soil remains buried (see Appendix B).

Decisions to retrieve buried waste or contaminated media generally rest on agreements among DOE, stakeholders, and regulatory agencies. Some or all of the waste buried at individual DOE sites may be left in place and monitored during long-term site stewardship programs (NRC, 2000c). However, DOE recognizes that some buried waste may require retrieval for treatment and disposition as TRU or mixed waste (DOE, 2001a).[11] Given the complex and changing nature of regulatory requirements and public concerns, the committee agrees that some buried wastes are likely to be retrieved in the future. Research begun now would be timely to address the additional challenges involved in locating and retrieving these materials.

---

[10]Retrievability is defined as the ability to remove waste from where it has been emplaced (IAEA, 1993).

[11]The TMFA Multi-Year Program Plan 2001 noted that the focus area was evaluating technology for ". . . automated retrieval of containerized waste that was not previously intended to be recovered." (DOE, 2001b, p. 47).

Baseline Technologies and Technology Gaps

DOE has no baselines for retrieving waste that has been buried at its sites throughout the country. However, initial plans to retrieve wastes at a small test plot at INEEL's Pit 9 provide an overview of state-of-the-art technology that is commercially available to DOE (see Sidebar 3.3).

At EPA Superfund sites, two of the most commonly used approaches for the retrieval of contaminated soil and groundwater are, respectively, excavating and removing soil and solid waste and pumping and treating contaminated groundwater. Soil is excavated using backhoes, bulldozers, or front-end loaders and placed on tarps or in containers. After excavation, the soil is removed by truck and taken to a licensed hazardous waste facility for treatment. Polluted water is extracted by pumping ground water into wells and up to the surface for placement in holding tanks. EPA also allows the treatment of waste in situ. For example, an oxidant is pumped into the ground to break down chemical contaminants. In situ oxidation is often faster than pumping and treating water in contaminated aquifers (EPA, 2002b; NRC, 1995, 1997b).

Retrieval of buried waste and contaminated media generally involves excavating the entire area where the material is known or expected to be. Extending this approach to the many acres that comprise DOE burial sites is probably impractical. There is a technology gap in locating and identifying specific objects (e.g., drums, gloveboxes) and determining if they need to be retrieved. Waste characterization is done as excavation proceeds, so surprises may be encountered.[12] An earlier project at Pit 9 led to concerns that drilling to retrieve waste samples could cause an explosion or fire (NRC, 2000a). Although robotics would be ideal for increasing worker safety, the current state of the art is that robotic systems lack the versatility and reliability needed for efficient deployment in the field (Sandia, 1998).

Challenges for Next-Generation Retrieval Technologies

From its fact-finding visits to DOE sites and committee members' experience and judgment, the committee believes that the greatest challenges for the next generation of waste retrieval technologies will be to provide

[12]A recent report on remediating a waste site at Sandia National Laboratories stated "the largely unknown characteristics of the buried waste material created uncertainties that could only be addressed during the excavation, rather than during the planning stages" (Methvin, 2002, p.1).

- improved, noninvasive means to locate and identify buried waste whether or not it is containerized;
- remote, noninvasive assessment of the condition of buried waste containers and potential leakage from the containers; and
- remote intelligent machines (robots) for waste retrieval and repackaging or treating as necessary.

Before the waste can be retrieved it must be located and at least a preliminary characterization must be made of its condition. If the drums or other containers are intact, they can be retrieved and handled using the processes developed for stored waste. Breached containers or non-containerized waste will be more difficult to retrieve. In either event, to minimize the number of processing steps and ensure worker safety, it would be helpful if more detailed characterization of the waste containers and their contents could be performed at the retrieval site. Hence there is opportunity to extend research for improved characterization methods, as described in the previous section, to the problem of buried waste. For example, research could lead to methods that are mobile, field deployable, and remotely operated. Next-generation technologies being developed for military purposes, such as land mine detection, might be adapted for locating buried waste (see Sidebar 3.4).

Physically retrieving wastes without exposing workers or spreading contamination will be a challenge. Next-generation robotic technology could be especially useful if the drums are not intact or if the soils surrounding the drums are highly contaminated. Robotic devices could repack the materials into a new container, preferably a smart drum capable of self-analysis and monitoring. Hanford proposed a remotely operated, multipurpose robotic vehicle with interchangeable actuators as one technology that would be capable of meeting a multitude of waste retrieval needs across the DOE complex. There is a specific need for robotic technology to retrieve RH-TRU wastes, some of which produce potentially lethal levels of radiation, from caissons located in the Hanford 618-11 waste burial grounds (Leary, 2002).

One of the best examples of next-generation robotics technology that might be further developed to retrieve buried waste drums is the HANDSS-55 system being assembled by the TMFA (see Sidebar 3.5). Drums are moved through the system automatically, and opened, and a robotic hand sorts through the contents of a drum to remove selected items as directed by video cameras and an operator using voice or touch screen commands (Frazee and Lott, 2002).

DOE has laid out an ambitious Robotics and Intelligent Machines Roadmap, which envisions threefold increases in productivity and tenfold reductions in radiation exposure to workers (Sandia, 1998). A previous NRC committee recommended robotics research for decom-

SIDEBAR 3.3 PLANS FOR BURIED WASTE RETRIEVAL AT INEEL PIT 9

INEEL's Radioactive Waste Management Complex (RWMC) was established in 1952 for disposal of solid low-level radioactive waste generated on-site. Wastes from other DOE sites were also buried there, including transuranic waste from Rocky Flats. Wastes were disposed in pits, trenches, soil vaults, an above-ground disposal pad, a transuranic storage area release site, and three septic tanks. One of the trenches contained in the complex is Pit 9, a 1-acre site that was used for waste disposal primarily from Rocky Flats between 1967 and 1969. DOE estimates that Pit 9 contains about 7,100 cubic meters (250,000 cubic feet) of sludge and solids contaminated with plutonium and americium.

An effort in the early 1990s to clean up Pit 9 failed, in part due to inadequate characterization of the wastes buried there. At the time the project was stopped, the cost of the cleanup was estimated to be about a half-billion dollars (GAO, 1997). INEEL has continued to develop options for Pit 9. INEEL's current planning is described here to provide examples of commercially available state-of-the-art technologies for waste retrieval.

The planned pilot-scale retrieval of wastes from portions of Pit 9 will probably be done in some type of containment structure. Modular-type structures are available that are inexpensive, easy to assemble, include their own air filtration systems, and can be assembled in basically any size and shape. Once assembled, the interior surface is sprayed with a strippable coating to prevent contamination of the structure itself. This coating can easily be removed or reapplied over the original coat to decontaminate or fix loose contamination during the life of the project.

Once the containment structure is ready, the next activities would be preliminary characterization and excavation. Excavator-mounted real-time systems are available to monitor radioactivity levels as the excavation proceeds. Automated systems are also available to assay the soil as it is removed.

Remotely operated excavation systems, such as the BROKK demolition machine (Holmhed Systems AB, Skelleftea, Sweden), which is equipped with a robotic arm, and the Sonsub Overburden soil removal system, are available for this phase of the work (see Figure 3.2). All of these technologies incorporate video systems to give operators visual information on all activities. Robotic sample collection capabilities are also available if needed or desired.

Once an object is encountered in the excavation and removed, a number of options are available for handling it. Following a remote inspection by video and radiation survey instruments, it will probably be repackaged for transportation to another facility. During the excavation and removal phase, a number of new fixative sprays and foggers are available to fix contamination and suppress airborne activity. These can be deployed remotely using the BROKK machine or other remotely operated equipment. Soft Sided Containers (Transport Plastics, Inc., Sweetwater, Tenn.), approved by DOT for shipping low-level

FIGURE 3.2 The BROKK demolition machine, equipped with a robotic arm, is an example of state-of-the-art technology that might be used to retrieve buried waste from Pit 9 at INEEL. Source: INEEL.

radioactive waste, are inexpensive and versatile for repackaging excavated objects or wastes that have somewhat irregular sizes and shapes. They can also be used to containerize both clean and contaminated soil associated with the retrieval activities.

For inspecting and characterizing the waste removed during retrieval operations, new instruments that identify radionuclides, such as the Surveillance and Measurement System (SAMS, Berkeley Nucleonics, San Rafael, Calif.); heavy metals, such as the Multi-element Spectrum Analyzer (NITON Corp., Bend, Ore.); and PCBs, such as the Spectro Xepos x-ray fluorescence analyzer (Asoma Spectro, Fitchburg, Mass.) are available for quick identification of contaminants and they eliminate the need for sample collection and laboratory work. The SAMS radiation detection system provides real-time isotope analysis in addition to radiation field strengths.

Following excavation of the desired materials, automated radiation survey systems—deployed either from the excavator or by other remotely operated devices—are planned to perform detailed surveys of the excavated pit to allow proper backfilling and monitoring.

SOURCE: R. Meservey, INEEL.

SIDEBAR 3.4 DEPARTMENT OF DEFENSE LAND MINE DETECTION

The DOD has identified 19 candidate technologies for land mine detection (GAO, 2001). The technologies include methods that exploit properties of the electromagnetic spectrum either passively (e.g., electromagnetic signature at infrared, millimeter wave, or microwave frequencies), actively by using electromagnetic energy (e.g., conductivity or resistivity, electromagnetic induction, electromagnetic radiography, gamma-ray imaging, LIDAR [light detection and ranging], microwave enhanced infrared, quadrupole resonance, radar, terahertz imaging, X-ray backscatter, X-ray fluorescence), or by other technologies (e.g., acoustic or seismic, biosensors, neutron activation analysis, trace vapor).

One example of a new mine detection technology is the timed-neutron moderation technique that uses neutrons to detect hydrogen in casings and explosives found in both plastic and metal land mines (Craig et al., 2000). A neutron source, about the size of a pager, holds a small amount of californium-252. As the element spontaneously fissions, it emits neutrons that electronics in the instrument then "time tag," noting when the decay occurred. The neutrons penetrate the soil, where they lose energy if they interact with hydrogen in a mine. These less energetic, slow neutrons are reflected back toward the detector. Helium-3 in low-pressure pipes collects the neutrons and emits electrons. The electronic signal is processed by special circuitry to indicate the potential presence of a land mine. The technique discriminates against other forms of hydrogen, such as in ambient moisture.

The DOD has also developed the Multi-sensor Towed Array Detection System (MTADS) for the underground imaging of metallic objects with particular emphasis on unexploded ordnance (Nelson and McDonald, 2001). MTADS consists of a low-magnetic-signature vehicle that is used to tow linear arrays of magnetometer and pulsed-induction sensors to conduct surveys of large areas. The MTADS magnetometers are cesium vapor full-field selected for low noise and intersensor reproducibility. Eight sensors are deployed as a magnetometer array on an aluminum and composite platform. The pulsed-induction sensors are deployed as an overlapping array of three sensors on a nonmetallic trailer. These sensors transmit a short electromagnetic pulse into the earth. Metallic objects interact with this transmitted field that induces secondary fields in the object. These secondary fields are detected by six detection coils that are located with and above the transmit coils.

The MTADS has demonstrated an impressive target location and depth prediction capability. Detection rates are greater than 95 percent under a variety of conditions. In addition to accurately locating a target for remediation, target classification is improved by developing increasingly detailed models of sensor response, focusing on the target shape information contained in the pulsed-induction response. Current projects are examining the potential of both frequency- and time-domain induction sensors for target characterization.

missioning nuclear facilities but cautioned that the DOE roadmap's "envisioned leaps in technology are not likely to occur without new knowledge" (NRC, 2001b, p. 66).

Microorganisms can have a profound impact on the chemistry and fate of buried waste (Newman and Banfield, 2002). Research in the public and private sectors has led to extensive knowledge of the bio-

SIDEBAR 3.5 HANDSS-55 REMOTE MECHANICAL SYSTEM

HANDSS-55 is a remote, partially automated, modular waste sorting and repackaging system for 55-gallon drums of contact-handled TM wastes. This system is designed to satisfy the unique and varying needs of each DOE site. Each module can be operated individually or in integration with the others. HANDSS-55 is being developed at the Savannah River Site, which has about 10,000 drums of Pu-238 and Pu-239 waste that must be handled in a contained facility for contamination control.

The HANDSS-55 system remotely opens 55-gallon drums and their polyethylene liners, gains access to the waste, removes items that are noncompliant for shipment to WIPP, and repackages the waste into polyethylene canisters. The used drums are shredded. Future plans include adapting the technology to a mobile platform as well as fully automating it.

SOURCE: Frazee and Lott, 2002.

geochemistry and fate of traditional pollutants in the environment, for example, the DOE Natural and Accelerated Bioremediation Research (NABIR) program.[13] The committee found no studies on the complex relationships among microbes and the organic and inorganic constituents within TM wastes themselves.

Research Opportunities

Prior to retrieval, it will be necessary to determine the condition of the waste or waste container. This need extends the challenge for imaging science research, described previously, to objects below ground. The approaches could be nonintrusive (preferred) or intrusive and could be coupled with chemical analysis. The nonintrusive approach may include ground penetrating radar, magnetometry, acoustics, chemical sensing of near-surface air samples, neutron activation, and radiological surveys. A minimally intrusive approach might use small-diameter boreholes to emplace equipment or sensors or to collect samples.

In addition to improving image resolution, research is needed on methods to improve object identification. Is it a drum, box, or rock? Is it intact? Is the soil surrounding the object contaminated? Are the contaminants stabilized or contained? Sophisticated image analysis and identification models and software will be needed to perform these assessments. The DOD's mine detection research might be leveraged (see Sidebar 3.4).

[13]See http://www.lbl.gov/NABIR/.

Research to understand biological processes that occur in buried waste can lead to better-informed decisions regarding retrieval. In simple experimental systems, the radiolytic effects of plutonium (primarily alpha-particle decay) have been shown to inhibit degradative or environmental microbes even at plutonium concentrations that do not cause chemical toxicity (Reed et al., 1999; Wildung and Garland, 1982). Understanding the relationships among waste materials and their associated microbial communities under real-world conditions (e.g., in soil, sludge, containers) could lead to improved predictability of the long-term fate and risk of the waste materials.

Significant advances in robotics will depend on research to make these devices more humanlike in their abilities to adapt to a variety of tasks, both physically and intellectually. Research toward more versatile actuators (the muscle of a robotic device), criteria-based software for independent decision making, and improved virtual reality systems for operators was recommended in a previous study of DOE facility decommissioning (NRC, 2001b). Such research would be equally relevant to developing retrieval technology for TM wastes.

Treatment

The EMSP should support research for treating TRU and mixed waste to facilitate disposal. This research should include processes to simplify or stabilize waste, with emphasis on improving metal separations, eliminating incinerator emissions, and enabling alternative organic destruction methods.

Treatment is defined as "operations intended to benefit safety and/or economy [of managing wastes] by changing the characteristics of the waste" (IAEA, 1993, p. 48). Treatment may be necessary to meet regulatory requirements. The results of treatment can include volume reduction, removal of radionuclides or other contaminants, and a change in the waste's composition.

TRU waste that meets shipping requirements can be sent to WIPP without treatment (see Chapter 2). According to DOE, after approval of Revision 19 to the Safety Analysis Report for Packaging (SARP), the volume of TRU waste that cannot be shipped due to gas generation has been reduced to approximately 3,000 cubic meters, or about 2 percent of the total inventory (Curl et al., 2002). Other shipping restrictions prohibit certain items in the waste and limit its heat production and its fissile material content.

Treatments for mixed low-level waste (MLLW) are prescribed in consent orders established between the sites and their host states in accord with the Federal Facilities Compliance Act of 1992 (FFCA; DOE, 2000d). The availability of landfills capable of accommodating MLLW, such as the Envirocare facility in Utah, has reduced the need for treatment of hazardous and radiological constituents. However, treatment needs remain for certain wastes, particularly those containing toxic constituents, which are subject to RCRA Land Disposal Restriction treatment standards. In addition to existing wastes, new MLLW will be generated through about 2070 (see Table 2.1).

Baseline Technologies and Technology Gaps

For TRU waste that does not meet shipping requirements, the baseline treatment is repackaging the waste. Repackaging may be necessary simply to remove prohibited items (see the previous section on "Characterization"). Repackaging waste to meet shipping requirements that govern heat production, fissile material content, or potential flammable gas production is extremely inefficient. According to the TMFA, repackaging these wastes so that they meet shipping regulations may result in a volume increase of ten- or perhaps twentyfold. In addition, about 98 percent of the TRU wastes that require remote handling will have to be repackaged. There is no baseline technology currently deployed for RH-TRU (Moody, 2002).

Baseline treatment technologies for MLLW developed at each DOE site as required by the FFCA were reviewed in a previous report (NRC, 1999a). Table 3.1 gives a summary of these treatment and stabilization options. Incineration is prominent among the options. However, incineration has been abandoned or is being phased out due to public concern about atmospheric emissions as well as site-specific economic considerations. As noted in Chapter 2, public opposition to a proposed incinerator at INEEL led DOE to create a Blue Ribbon Panel to recommend alternatives to incineration. A brief history of incineration and its alternatives is given in Appendix C.

The TMFA Multi-Year Program Plan states that three to five primary alternatives to incineration will be selected for comparison testing at DOE's Western Environmental Treatment Office (WETO) in Butte, Montana, in fiscal year 2002 (DOE, 2001b). The current strategy is to select processes to represent the three general classes of alternatives: (1) thermal, (2) aqueous-based chemical oxidation, and (3) chemical separations. In addition to testing the primary alternatives at WETO, tests of other alternative methods at other locations will be conducted in a manner to make them consistent with the studies at WETO. Examples include the testing of a mediated electrochemical oxidation process at the DOD's Aberdeen facility for chemical warfare agents, a solvent extraction method at Florida International University, and a

TABLE 3.1 Summary of Treatment and Waste Form Options for Mixed Waste Groups

Waste Group	Hazardous Characteristics	Typical Hazardous Components
Wastewater (<1% organic)	Corrosive, toxic	Cr, Pb, Cd, Hg
Combustible organics	Ignitable, corrosive, toxic	Halogenated and non-halogenated solvents; Cr, Cd, Pb, Hg, PCBs
Inorganic, homogeneous solids and soils (<60-mm particles)	Toxic	Electroplating waste, solvents, Pb, Cr, Cd
Debris (>60-mm pieces)	Toxic	Pb, solvents
Unique	Ignitable, reactive, toxic	Reactive metals, compressed gases, explosives

SOURCE: DOE, 1997.

molten aluminum process at Sandia National Laboratories. The TMFA is currently developing guidebooks to assist DOE and permit writers in developing permit conditions for each of these alternative technologies.

Macro- and microencapsulation have become important baseline technologies for waste stabilization, i.e., treatment to prepare wastes for further handling or disposal (see Figure 3.3). Stabilization of mixed waste for disposal usually relies on its incorporation into one of several matrices—grouts or cements, glass, polymer, or ceramic—to produce a relatively homogeneous waste form, although some wastes are simply compacted.[14] Macroencapsulation yields a heterogeneous waste form

[14]Matrices for stabilizing TM wastes ("waste forms") were assessed by a previous NRC committee (NRC, 1999a).

Treatment Goal	Treatment Options	Available Waste Forms
Volume reduction, organic removal	Incineration; traditional water treatments: reverse osmosis, neutralization, precipitation; no treatment	Grout, polymer, glass, Hg amalgamation
Destroy organics, volume reduction	Incineration, thermal oxidation	Grout, polymer, glass, Hg amalgamation
Volume reduction, meet disposal requirements	Incineration, thermal oxidation, no treatment	Grout, polymer, glass, sulfur cement
Volume reduction, meet disposal requirements	Extraction: physical, chemical, thermal Destruction: thermal, biological, chemical Immobilization: microencapsulation, macroencapsulation, sealing	Grout, polymer, glass, Hg amalgamation, direct disposal of object or compacted material
Hazard reduction	Specific treatments for individual wastes or waste steams	Grout, polymer, glass, Hg amalgamation, direct disposal

by encasing the waste in a coating or block of suitable matrix, usually low- or high-density polyethylene or cement. Microencapsulation is used for the stabilization of ashes, salts, or other dry powders by mixing the waste with polymer (chiefly low density polyethylene) as feed for the extruder to produce pellets of intimately mixed waste and matrix. Versions of these technologies are used to stabilize approximately 20 percent of the waste requiring treatment for disposal at Envirocare, Utah.

Macro- and microencapsulated wastes are relatively robust mechanically when encased in a structurally rigid secondary container, and polyethylene is relatively inert to radiolysis at the levels of activity typically associated with MLLW. However, contaminants are not chemically fixed in this form, merely encased, so constituents may be more susceptible to leaching under scenarios of mechanical intrusion. For example, the EPA's Toxicity Characteristic Leaching Procedure requires grinding

FIGURE 3.3 Macro-encapsulation is used to stabilize heterogenous waste or large objects for disposal. Typically, grout or a polymer is poured over the waste so that it is encased physically. There are few data on the long-term durability of macroencapsulated wastes. Source: DOE Richland Operations Office.

the waste form if necessary to meet size criteria for the test. This can alter the barrier provided by the encapsulation. There are few data on the long-term durability of macroencapsulated wastes.

Thermal desorption is a relatively mature technology that can remove volatile organic compounds from solid TM wastes. Commercial units are available from several vendors, (e.g., Permafix Environmental Services, Sepradyne, Envirocare). The process is being used at Oak Ridge, and the TMFA has funded process development work for treating soils and sludges at Fernald, Ohio, and organic sludges at INEEL. To drive off VOCs and moisture, waste is heated to the range of 300 to 1200°F depending on the organics to be removed, the nature of the waste, and process details. Operating the process under reduced pressure (vacuum thermal desorption) allows lower temperatures to be used. An inert gas such as nitrogen can be used to purge the organics and prevent accidental ignition. Gases that are released are usually condensed or trapped, for example, on carbon. Thermal desorption is among the three leading options recommended as alternatives to incineration by the Blue Ribbon Panel, although the technology does not apply to all types of TM waste or reduce the waste volume (see Appendix C).

The TMFA recognized gaps in baseline technologies for treating small volumes of unique or problematic wastes. These problematic wastes include reactive materials, gas cylinders, and tritium-contaminated materials. Because their disposition requires specialized or one-of-a-kind approaches, they are often not economically attractive for private sector treatment contracts. Their limited quantities and special problems have kept them in relatively low priority for disposition at most sites. However, because they comprise about 10 percent of DOE's total TM waste inventory, they represent a potential roadblock to site closure. A Waste Elimination Team formed by the TMFA was involved in defining the inventory of these wastes and developing lists of technology needs (Hulet, 2002).

One example of problematic waste is mercury, and technology gaps remain for its treatment and stabilization. Mercury is present in a broad range of concentrations in several of DOE's mixed waste streams, including large volumes of soil and debris and several types of process residues (see Table 2.3). Because it is mobile and easily vaporized, the presence of mercury creates additional effluent monitoring and control concerns in incineration and can reduce the efficiency of MLLW stabilization processes. Removing mercury before treatment simplifies downstream treatment operations.

Depending on the concentration or form of the mercury, current EPA standards require stabilizing the waste in a form that passes the TCLP test, or else treating the waste by thermal desorption or retorting, which creates a separate waste stream that cannot be recycled and will itself require stabilization. The proposed methods of stabilization are amalgamation for elemental mercury and chemical immobilization through precipitation or sorption. Some stabilizing agents are based on sulfur, whereas others have proprietary formulations. One particular matrix, which has been evaluated recently, is sulfur-containing cement. Despite the fact that these and related methods have been under investigation for several years, there still do not appear to be robust baseline methods for treatment of all mercury-containing waste streams (Morris et al., 2002; Townsend, 2001).

The lack of application of biotechnologies in TM waste treatments appears to be a technology gap.[15] In the last 15 years, significant advancements have been made in biological treatment technologies for organic pollutants including chlorinated and aromatic solvents, cutting oils, PCBs, and related materials (Unterman et al., 2000). In many cases, biological treatment can result in a significant reduction in treatment costs over traditional technologies. Biological treatment approaches are sometimes coupled to other chemical or physical methods (e.g., air sparging and vapor extraction, carbon adsorption, washing) to reduce overall treatment costs. Advances in bioreactor design to allow efficient and safe treatment of contaminants have accompanied advancements in biotreatment technologies (Fortin and Desshuses, 1999; Steffan et al.,

---

[15]The TMFA phased out a project for biodegradation of tritiated waste at Lawrence Berkeley National Laboratory (Maio and Reese, 2002).

2000). Although DOE's NABIR program has focused on basic research related to the immobilization or removal of heavy and radioactive metals from contaminated environments, little work has been done to evaluate biological leaching or removal of these materials from TRU and mixed wastes.

Challenges for Next-Generation Treatment Technologies

Based on the committee's fact-finding visits to DOE sites, recent approaches to developing alternatives to incineration, and committee members' own expertise, the committee believes that the greatest challenges for next-generation treatment technologies lie in developing

- emission-free treatment processes,
- treatments for problematic or unique wastes, and
- methods to ensure the long-term durability of stabilized waste.

The committee reviewed the recommendations for next-generation alternatives to incineration by the Blue Ribbon Panel and the programs initiated by the TMFA (see Appendix C). While agreeing that the recommended technologies show promise, the committee believes that any large-scale treatment processes are likely to meet with similar public concern as incineration unless more complete knowledge can be demonstrated regarding the formation of unwanted by-products, especially after process upsets. Further, concerned citizens should be involved in selecting among technological alternatives (see Chapter 2).

Plasma arc technology is a robust technology that can treat a wide variety of wastes, although it is not emission free. This technology, first investigated by the TMFA in 1996, has continued to mature. The Naval Research Laboratory (NRL) is managing and supporting a project to establish a plasma arc hazardous waste treatment system at the Norfolk, Virginia, naval base. This system will be capable of destroying most of the 2.5 million pounds of hazardous waste generated annually at the base. A plasma arc research facility at NRL is also being used to support a Navy Advanced Technology Demonstration Project to develop a pre-prototype shipboard plasma arc system for destroying solid waste onboard Navy ships. The EMSP is supporting a plasma torch technology for decontaminating DOE facilities (NRC, 2001b).

Recently, the TMFA has chosen AEA Technology Engineering Service's "Silver II" method for further testing.[16] The process uses Ag^{2+}

---

[16]See http://www.sciencedaily.com/releases/2002/01/020104074240.htm and http://www.aeat-prodsys.com/prodsys/divisions/OCD.html.

and concentrated nitric acid to oxidize organics, followed by electrochemical regeneration of the Ag^{2+} and recovery of the nitrogen oxides. The Silver II method produces essentially no emissions but treats a smaller spectrum of wastes than incineration or plasma processes. Silver II operates at low temperature, is easy to control, treats many organic wastes, reduces waste volume, produces no dioxins, and does not require pretreatment for small solids, slurries, or liquid wastes. However, the pretreatment of larger solid organic wastes may be required. The process is being evaluated as an option for destroying organics in Pu-238 waste at the Savannah River Site (Pierce, 2001). The U.S. Army is testing Silver II at the Aberdeen Proving Ground to destroy chemical weapons agents.

Low-emission combustion technologies are used at many of today's petrochemical refineries. These systems (called enclosed zero flares, ground flares, population area combustors, or thermal oxidizer flares) control emission from smoke stacks, resulting in no smoke, odor, or objectionable noise.[17] Such technology might provide a starting point for developing near-zero emission technologies for the more complex challenge of burning TM wastes.

The current state of the art in metal ion separation is the use of specific liquid-phase extractants or of solid-state sorbents or ion-exchange materials capable of achieving specificity for metal ion removal, particularly from aqueous waste streams. These technologies had their origin in the development of process chemistry for actinide purification and mining operations. In recent years, the emphasis in the design of such extractants shifted from increasing efficiency for producing nuclear materials to increasing separation factors for waste stream polishing (NRC, 2000b). More recently, research has begun to focus on the design of new separation systems, such as ion-specific membranes, in a desire to minimize secondary waste streams traditionally associated with liquid processing schemes.

In addition to treating legacy wastes, an important role and challenge for next-generation separation processes will be improved product separations for DOE's continuing mission to produce nuclear materials. During its visit to Oak Ridge's isotope production facility the committee was reminded that greater efficiency in separating highly radioactive products means a less radioactive and easier-to-manage waste stream.

Understanding factors that affect durability of matrices for mixed wastes disposed in near-surface, RCRA-compliant landfills will be especially important as DOE moves its MLLW from storage to disposal in site closures. Current testing protocols such as the TCLP are designed

[17]See http://www.johnzink.com/markets/html/m_hpi.htm.

for homogeneous waste forms, in which the waste is incorporated into the matrix (e.g., sludges in grout or glass). These current tests may not be good predictors of susceptibility to leaching under conditions different from those specified in the protocol (see the discussion of predictive modeling). As DOE and its subcontractors seek to reduce the costs of treating MLLW by greater use of encapsulation, developing methods to ensure the long-term durability of these heterogeneous waste forms will become increasingly important. As noted earlier, there are essentially no durability data for micro- or macroencapsulated wastes.

Research Opportunities

There are research opportunities in areas of chemical treatments (including advanced alternatives to incineration), biological treatments, stabilization, and waste form durability.

Chemical Treatment

Essential to developing publicly acceptable alternatives to incineration is research to develop sensitive, reliable, and practical detection methods to track both radionuclide and hazardous chemical materials during the treatment processes. Answering the question, How much hazardous material is being released to the environment? is particularly important for public acceptance of treatment methodologies. It is also essential for controlling potential risks to workers during treatment operations.

Research opportunities for treating TM wastes are in accord with opportunities reported in a recent study of the technical needs for the Deactivation and Decommissioning Focus Area (NRC, 2001b). In particular, research in the speciation of inorganic constituents in wastes may have an impact on the selection of future treatment options. The state of the art in examining metal ion speciation is much more developed for actinide constituents in wastes than for other inorganic constituents.[18] Further research is needed to assess the chemical forms of other metals (e.g., oxidation state, speciation) in the presence of complex matrices and a variety of co-contaminants. More research is needed to understand the nature of the oxide-substrate interaction. With the advent of new molecular design tools associated with nanotechnology applications, a wealth of new opportunities exists to generate novel

---

[18]See *MRS Bulletin,* Vol. 26, No. 9, September 2001.

separation schemes based on the design of porous materials with chemical functionality specific to the separation tasks at hand.

Research can improve methods to dissolve plutonium oxide selectively in the presence of other metals and organics or lead to methods of controlling high-activity, finely divided particulate materials. For example, the Pu-238 isotope in waste from plutonium processing at the Savannah River Site exists as very finely divided oxide powder contaminating heterogeneous wastes. Due to the wattage restrictions on waste to be packaged for WIPP, it is unclear whether this waste will meet DOT requirements. The general practice for treating mixed waste would be to destroy the organic constituents through incineration or an alternative technology. The danger of dispersal of the powder precludes most thermal treatments, however, suggesting the need for a new method for removing and stabilizing the oxide powder. This raises several interesting technical challenges, such as ensuring efficient removal of finely divided powders, controlling particle dispersion and perhaps inducing agglomeration, and avoiding the generation of additional waste streams.

Hydrogen generation remains a factor in the shipment of containerized waste to WIPP. The TMFA actively sponsored work on hydrogen getters (absorbers). During a visit to the Savannah River Site, the committee heard a presentation describing significant advancements made toward the development of polymeric hydrogen getters to capture the hydrogen produced in waste drums (Duffey, 2001). These getters have proven useful for most wastes tested to date, but poisoning of the polymers—evidently by organic vapors—can occur, thereby reducing their efficiency. There are research opportunities for understanding the fundamental processes of both hydrogen production and hydrogen adsorption.

Biological Treatment

There are also opportunities to develop efficient and cost-effective biological treatment technologies for TM wastes, including recovered soils and sediments (see Table 2.1). Research should focus on the treatment of readily degradable organic material and the combined application of biological and physical or chemical treatment technologies. Although physical methods such as steam stripping, heating, or vacuum extraction may be the most common for separating the various waste components, biological treatment may be appropriate for some waste types. For example, sludges containing water and biologically degradable organic compounds may be amenable to biological treatment for removal of the hazardous components.

Research opportunities include the identification of enzyme or whole-cell treatment approaches that target specific contaminants (e.g., PCBs, mercury) or broad categories of contaminants (e.g., combustible solvents, cutting oils, chlorinated hydrocarbons). Treatment technologies must be amenable to application in the environment of the containerized waste. For example, enzyme systems that function at extreme pH levels, high salinity, or under high solvent or nonaqueous conditions would be desirable. Application of advanced molecular techniques such as directed evolution (Stemmer, 1994) might help develop appropriate biocatalysts. Research directed toward identifying ways to apply the biocatalysts, such as novel immobilization matrices, is also necessary to facilitate successful use of these catalysts.

Fundamental research may identify biological treatment processes that can facilitate the removal of RCRA wastes (e.g., Hg, Pb) and radioactive metals from contaminated media. Like hazardous organic chemicals, some toxic metals are susceptible to biological transformations. For example, mercuric ion (Hg^{2+}) can undergo a range of biological transformation processes including reduction to elemental mercury, which is volatile at room temperature, or methylation, which forms the highly toxic monomethylmercury or the volatile dimethylmercury. These biotransformations have shown promise for removing mercury from wastewater (Wagner-Dobler et al., 2000).

Recent research has shown that some common iron-reducing soil bacteria can solubilize plutonium hydrous oxides that bind tightly to soils (Rusin et al., 1994). Adding a chelator enhances the solubilization process. These findings suggest that biological treatment, via either bioaugmentation or biostimulation, coupled with soil washing technologies could provide a mechanism to remediate actinide-contaminated soils. Similarly the common soil microbe *Microbacterium flavescens* can absorb and accumulate Pu(IV) if the siderophore desferrioxamine-B is provided (John et al., 2001).

Siderophores are chelators that are produced and released by many iron-utilizing bacteria in soil environments. The siderophores bind iron, and the iron-siderophore complex is captured by iron-utilizing microbes. The fact that these iron chelator compounds can also bind actinides suggests that they can be exploited to treat TM waste. Bioremediation approaches could involve either stimulating siderophore production by indigenous organisms or adding exogenous siderophore-producing organisms or siderophore-containing extracts to the waste or contaminated media. Once chelated, the soluble actinides could potentially be removed by soil washing or related methods.

Research is needed to develop reliable processes to transform or remove heavy and radioactive metals from mixed waste. The research should be based on the large existing volume of research on heavy-metal biotransformation and should evaluate the effects of multiple contaminants, radioactivity, and extreme environmental conditions (e.g., low or high pH, high salts, cementing agents) on metal mobilization, immobilization, accumulation, speciation, and related transformations. The research should include multidisciplinary studies combining microbiology and genetic engineering, materials handling, reactor design, and process engineering to develop cost-effective technologies for treating TM wastes.

Biotechnology research opportunities include the identification and development of improved hydrogen and methane scavengers that can be added to waste drums. Potential scavengers could include efficient hydrogen- or methane-scavenging microorganisms or enzymes capable of binding or oxidizing hydrogen or methane. The enzymes could be improved by using genetic engineering and directed evolution of enzymes (Stemmer, 1994) to enhance their hydrogen- and methane-scavenging efficiency. Additional research could include identification or development of improved methods for applying such organisms or enzymes to the drums. This could include development of immobilization techniques (e.g., sol gels, polyurethane) that improve activity and allow long-term survival of the biocatalyst while not exceeding the free liquid limits imposed by waste disposal facilities. This also could lead to studies to identify artificial electron donors that could be added to hydrogen and methane oxidation enzyme preparations to maintain their oxidative activity over long periods. Enzymes such as methane monooxygenase also may destroy other compounds such as carbon tetrachloride that poison some chemical hydrogen scavengers.

Stabilization

Retrieval of buried waste or contaminated media has generally required the use of very expensive engineering controls to ensure the safety of workers and the surrounding environment. A tremendous reduction in costs could be realized if waste were stabilized prior to or in the early stages of retrieval. Current containment methods involve either application of simple barriers or, in some cases, methods that partially stabilize the waste (e.g., grouting). Research should address new systems for stabilization. One could envision the development of smart materials that react with waste constituents to generate optimized coatings or combined chemical and biological processes that would stabilize the waste selectively by alteration of the matrix or generation of additional barrier layers. A recent paper describes a smart Portland cement that senses environmental conditions (Wen and Chung, 2001).

A previous report recommended research for stabilizing or containing contamination in soils or ground water (NRC, 2000a). There are important distinctions, however, between technology needs associated with the containment of subsurface contaminants in a particular geological environment and the stabilization of buried waste or contaminated media to facilitate retrieval. The need to control subsurface contaminants suggests research to devise barriers that function over the time scale envisioned for long-term site stewardship, whereas stabilization of buried waste may require only interim containment until such time as the waste is emplaced in a disposal facility.

A need exists for fundamental proof-of-concept investigations to determine the potential for microbial processes to stabilize buried TM waste by altering its composition. Promising biocatalysts may be obtained by applying traditional microbial selection and enrichment approaches to target waste materials or by "biomining" other radioactive waste materials to identify promising radiation-resistant degradative microbes. Genetic engineering could be applied to improve the metabolic capabilities of radiation-resistant organisms to develop improved biocatalysts for treating mixed waste (Brim et al., 2000; Lange et al., 1998). Organisms also can be developed to function in the extreme environmental conditions (e.g., high salt, extreme pH) found in some waste types. Research on microbiology should be coupled with research in chemistry, materials science, and process and reactor engineering to develop integrated systems to handle and treat difficult waste materials.

Waste Form Durability

The NRC study of waste forms found that the matrices (e.g., grout, glass, polymers) available to stabilize MLLW for disposal are adequate (NRC, 1999a). However, the report noted that most repository performance assessments do not take credit for waste forms because quantitative tests for their long-term durability have not been developed. The report recommended that DOE's Office of Science and Technology (OST) support work aimed at fundamental understanding of waste form durability and suggested that EMSP evaluate and fund proposed research in this area. The committee agrees that this is a valuable area for research.

Research to understand the chemical and physical processes that may leach or degrade waste forms in RCRA-compliant landfills should be combined to develop predictive models and more appropriate test procedures. As noted earlier, very little is know about the long-term durability of heterogeneous waste forms or microbiological effects on waste forms. A specific opportunity would be evaluation of the long-term stability of, and effects of biological activity on, metals immobilized in sulfur cements and related materials. Bacteria are able to metabolize sulfur compounds in a variety of ways, including oxidation of reduced species (which can form sulfuric acid) and reduction of oxidized species, depending on the organisms present and the redox potential. Understanding how these reactions affect the leachability of RCRA metals from these new immobilization materials is needed.

Long-Term Monitoring

The EMSP should support research to improve long-term monitoring of stored and disposed TRU and mixed wastes. Research should emphasize remote methods that will help verify that the storage or disposal facility works as intended over the long term, provide data for improved waste isolation systems, and inform stewardship decisions.

Long-term monitoring will play an important role in many of DOE's site cleanup activities, especially in continuing stewardship of the sites (NRC, 2000c). In the context of TM waste, long-term monitoring will be important both during storage and after the waste has been emplaced in a disposal facility. The storage phase is likely to be long for some wastes, and it will be necessary to ascertain that the containers maintain their integrity and that internal processes in the containers are as predicted (see the section on "Characterization"). Prudence requires that wastes disposed in a RCRA-type landfill or in a repository such as WIPP be monitored until the facility is closed. Post-closure monitoring of waste in RCRA landfills will be important to ensure continuing safety and to provide data for maintaining the landfill.

In providing advice on ensuring the long-term safety of WIPP, a previous NRC committee found that ". . . the activity that would best enhance confidence in the safe and long-term performance of the repository is to monitor critical performance parameters during the long pre-closure phase of repository operations (35 to possibly 100 years)" (NRC, 2001d, p.1). For example, chambers to hold waste in the WIPP will be excavated as needed, drums will be emplaced, and lithostatic forces gradually will close the chambers—crushing the containers and sealing the waste in place. Early emplacements will be sealed during the operational life of the WIPP. Monitoring the waste during this sealing process can yield important scientific understanding of the actual closure process and also enhance safety by determining if the closure occurred properly.

Baseline Technologies and Technology Gaps

In its review of DOE's environmental quality portfolio, the Strategic Laboratory Council found a significant gap in the "fate and transport of contaminants and performance monitoring to support waste repositories and inform stewardship decisions" (DOE, 2000c, p. iii). DOE has no established plan for monitoring its MLLW disposal facilities (e.g., Hanford, the Nevada Test Site) beyond the 30-year RCRA compliance period. There are no plans for mechanical or chemical monitoring of closed rooms in WIPP during its operating period or thereafter.

Understanding and verifying the long-term behavior of these wastes and their disposition systems require more attention to monitoring than is apparent in DOE's current planning. Research begun now to develop scientifically sound, simpler, and more reliable technologies for long-term monitoring can help ensure the safety of stored or disposed TM waste, reassure concerned citizens, provide data for new disposal facilities in the United States or abroad, and contribute generally to scientific knowledge.

Challenges for Next-Generation Technologies

Based on its fact-finding visits to DOE sites and committee members' own expertise and judgment, the committee believes that future challenges for long-term monitoring technologies will be to extend the next-generation technologies described in the section on characterization to enhance reliability, stability, and remote operation, including

- long-lived, reliable sensors (and power supplies) that can be remotely interrogated, and
- airborne or satellite imaging.

The sensor technology discussed previously is generally applicable for long-term monitoring. State-of-the-art improvements in technology are making it possible to interrogate sensors from remote locations and to provide remote, standoff detection of both chemical and radiological hazards. Distributed sensors can be monitored by Internet and wireless technologies (Pottie and Kaiser, 2000).

In one example of a next-generation sensor technology, an electric utility prevents the overheating and shutdown of its power grid by monitoring a network of transformers and nodal sites with distributed sensors and the Internet.[19] This replaces the expensive patchwork of

[19]See http://www.graviton.com/.

wired networks and human intervention required by present technologies to keep the grid in operation. The wireless sensors used in the new technology work above or below ground, and their spread-spectrum signal is impervious to electromagnetic interference. If the temperature of a transformer exceeds a safe level, the sensors trigger an alarm that alerts the utility. The utility reroutes electricity around the problem area or shuts down affected areas. By avoiding transformer failures, the utility is able to prevent costly shutdowns and blackouts. Sensors to measure other parameters can be implemented readily within the system's architecture.

Another example includes the monitoring of mobile platforms. Transportation of materials on trucks, trains, ships, planes, and buses produces an array of potential data management system needs. These needs include environmental and safety monitoring, preventive maintenance, global positioning, antitheft measures, and real-time engine sensor data. Platform mobility and the lack of cost-effective wireless connections have been the limiting factors in developing sensor systems for mobile platform services. Again, the data can be managed through wireless sensor networks for mobile platforms. The internal temperature of a compartment can be measured to ensure that the temperature stays within range, video monitors can be utilized for theft prevention and for ensuring compliance with safety regulations, and noxious fumes can be detected to allow for timely correction. Sensor data are transmitted by a wireless wide-area network connection. Clients on mobile platforms gain cost-effective access to data about location, safety, security, engine stability, inventory, and many other parameters that can be accessed in the field or from corporate command centers.

All materials and objects (e.g., soil, water, trees, vegetation, structures, metals, paints, fabrics) create a unique spectral fingerprint. An optical sensor can determine these fingerprints by measuring reflected light, most of which registers in wavelengths, or bands, invisible to the human eye. Commercial state-of-the-art hyperspectral imaging systems operate across up to 220 wavelengths to record precise images of an otherwise hidden world. Where a standard sensor with fewer than 10 bands is capable of differentiating only between gross classes of vegetation, a hyperspectral imager can discriminate between plants and is sensitive enough to separate healthy from unhealthy growth. Hyperspectral sensors and imaging offer many attractive features for long-term, remote monitoring applications.[20] Current and advanced technologies include the following:

[20]See http://www.techexpo.com/WWW/opto-knowledge/hyperspectrum/index.html.

- remote sensing of earth resources,
- chemical detection and cloud tracking,
- medical photodiagnosis, and
- positional radionuclide concentrations.

A better way to monitor large, remote sites may be from airborne or satellite platforms.[21] Overflight monitoring of nuclear power plants and other nuclear facilities for radionuclide emission is commonly practiced today. Mineral and oil prospecting is also done from the air.

Research Opportunities

There are research opportunities for developing remote, distributed sensor systems to achieve self-monitoring "smart" drums and in monitoring TM repositories to achieve more fundamental knowledge of physical, chemical, and biological processes that govern their behavior.

Remote Sensing

Smart sensors can dramatically improve the monitoring of waste storage and waste disposal in near-surface (RCRA) landfills by creating smart drums that self-analyze and report their content and location. Smart filters could monitor and control (i.e., vent or getter) the gas produced in the drum. The Internet and wireless technologies can monitor these and other distributed sensors remotely during all phases of waste disposition, including storage, transportation, and disposal. The availability of inexpensive and reliable sensors for chemical and radiological hazards in the drums would also be beneficial. To be cost-effective, the sensors must be small and mass produced like today's MicroElectroMechanical Systems (MEMS) and tomorrow's Nano-ElectroMechanical Systems (NEMS). To be practical, the sensors must be self-sufficient, harvesting their energy from the environment or from the waste itself.

Research on hyperspectral-imaging methods coupled with image-processing algorithms could lead to advanced remote-sensing technologies that would be well suited to monitor large, remote sites from airborne or satellite platforms.

---

[21]See, for example, the *International Journal of Remote Sensing*.

Repository Behavior

If WIPP is used only as a geological repository for the disposal of TRU wastes, then a scientific opportunity to advance our understanding of this unique major facility will be missed. In addition to being a repository, WIPP can be an important laboratory for geoscience and sensor technology. Scientific research could be done in conjunction with measures to ensure WIPP's long-term safety. The results will be indispensable if WIPP is enlarged or if a new salt repository is needed. A previous NRC WIPP study recommended pre-closure monitoring to gain information on the following:

- room deformation, healing of the disturbed zone around the rooms, and performance of shaft seals;
- brine migration and moisture access to the repository;
- gas generation rates and volumes; and
- effectiveness of materials placed around the waste (e.g., MgO) to modify its chemical environment (NRC, 2001d).

Sensors developed and tested at WIPP could then be used for long-term monitoring of other repositories, landfills, and burial sites. These sensors must be robust and have a lifetime of at least 10 years to monitor room closure. They should be controlled remotely and monitored by the Internet and wireless technologies, and they must be self-sufficient, harvesting their energy from the salt or the waste itself.

Microbes are likely to exist and evolve in wastes in the WIPP and in RCRA landfills. A better understanding of these microbes and their activities will help predict the long-term fate of the different waste forms and their components. Microbial activity may destroy or immobilize some waste components while increasing the motility or toxicity of others. Research should focus on specific processes, including biodegradation of the various organic components of the waste and reactions altering the geochemistry of the inorganic components. Research should evaluate biogeochemical factors that can affect the leaching or migration of toxic and radioactive materials in the environment and the effect of physical conditions (e.g., pH, temperature) and chemical composition (e.g., organic and inorganic components, oxygen or other electron acceptor availability) on biogeochemical processes occurring in the waste.

Near-Term and Longer-Term Research

Accelerating site closure, a key feature of Office of Environmental Management (EM) planning since the 1990s, has been emphasized by

EM's recent top-to-bottom review. Among the areas for EMSP research recommended by the committee, research in characterization that would expedite shipping wastes for off-site disposal is most likely to provide immediate payoffs. Research toward methods for treating wastes that do not meet shipping or disposal criteria might provide similar near-term payoffs.

Nevertheless, closing the larger DOE sites will require decades. Problems that are not foreseen or appreciated today are likely to be encountered in buried waste retrievals. Monitoring the WIPP during its operational period is a unique scientific opportunity. Demonstration that WIPP behaves as expected could be invaluable as DOE seeks to open other geological waste repositories. These opportunities for the long-term, breakthrough research envisioned by Congress should not be overlooked in the rush to meet short-term needs.

References

Anderson, A.W., H.C. Nordon, R.F. Cain, G. Parrish, and D. Duggan. 1956. Studies on a radio-resistant micrococcus. I. Isolation, morphology, cultural characteristics, and resistance to gamma radiation. Food Technology 10:575-578.

Battista, J.R. 1997. Against all odds: The survival strategies of *Deinococcus radiodurans.* Annual Review of Microbiology 51:203-224.

Brim, H., S.C. McFarlan, J.K. Fredrickson, K.W. Minton, M. Zhai, L.P. Wackett, and M.J. Daly. 2000. Engineering *Deinococcus radiodurans* for metal remediation in radioactive mixed waste environments. Nature Biotechnology 18(1):85-90.

Brockman, F.J. 1995. Nucleic-acid-based methods for monitoring the performance of in situ bioremediation. Molecular Ecology 4:567-578.

Busenberg, G.J. 1999. Collaborative and adversarial analysis in environmental policy. Policy Sciences 32(1):1-11.

Case, D.R. 1991. Resource Conservation and Recovery Act. Pp. 406-442 in Environmental Law Handbook, 11th ed., J.G. Arbuckle et al., eds. Rockville, Md.: Government Institutes.

Chopyak, J., and P. Levesque. 2002. Public participation in science and technology decision making: Trends for the future. Technology in Society 24(1-2):155-56.

Cohn, J.P. 2002. Environmental conflict resolution. Bioscience 52(5):400.

Collins, G.E., and Q. Lin. 2001. Radionuclide and metal ion detection on a capillary electrophoresis microchip using LED absorbance detection. Sensors and Actuators B: Chemical 76(1-3):244-249.

Collins, G.E., Q. Lin, S. Abubeker, and E. Vajs. 2002. Remote fiber-optic flow cell for the detection of uranium(VI) in groundwater. Applied Spectroscopy 56(4):464-468.

Craig, R.A., A.J. Peurrung, and D.C. Stromswold. 2000. Mine Detection Using Timed Neutron Moderation. The Unexploded Ordinance (UXO)/Countermine Forum, April. Available at: <http://www.pnl.gov/nsd/commercial/tnd/Images/UXOconf2a%20.PDF>.

Curl, R., W. St. Michel, and S. Lott. 2002. Payload enhancement for transporting TRU waste within restrictive regulatory limits. Presented to the TRU and Mixed Waste Focus Area End-User Review, Baltimore, Md., March 19-21. Available at: <http://tmfa.inel.gov/Documents/MY02Trans.pdf>.

DOE (U.S. Department of Energy). 1995. The DOE National 1995 Mixed Waste Inventory Report. Washington, D.C.: U.S. Department of Energy.

DOE. 1997. MWFA Technical Baseline Report. DOE/ID-10524. Vols. 1 and 2. April. Idaho Falls, Idaho: U.S. Department of Energy.

DOE. 2000a. Environmental Management Science Program. Annual Report. FY 2000. Office of Science and Technology. DOE/EM-0569. Washington, D.C.: U.S. Department of Energy.

DOE. 2000b. Report of the Secretary of Energy Advisory Board's Blue Ribbon Panel on Emerging Technological Alternatives to Incineration. Washington, D.C.: U.S. Department of Energy.

DOE. 2000c. Adequacy Analysis of the Environmental Quality Research and Development Portfolio. Draft revision, July. Washington, D.C.: U.S. Department of Energy.

DOE. 2000d. Transuranic and Mixed Waste Focus Area Annual Report. FY 2000. Washington, D.C.: U.S. Department of Energy.

DOE. 2000e. Management Plan. Environmental Management, Office of Science and Technology. Research and Development Program. September. Washington, D.C.: U.S. Department of Energy.

DOE. 2001a. Summary Data on the Radioactive Waste, Spent Nuclear Fuel, and Contaminated Media Managed by the U.S. Department of Energy. April. Washington, D.C.: U.S. Department of Energy.

DOE. 2001b. Transuranic and Mixed Waste Focus Area Multi-year Program Plan. DOE/ID-10659. April. Washington, D.C.: U.S. Department of Energy.

DOE. 2001c. Action Plan for Emerging Technological Alternatives to Incineration. Environmental Management Office of Integration Disposition. June. Washington, D.C.: U.S. Department of Energy.

DOE. 2001d. Transuranic Waste Characterization Cost Analysis. U.S. Department of Energy Office of Environmental Management. Memorandum from L. Wade, Director of the Waste Isolation Pilot Plant Office. February 23. Washington, D.C.

DOE. 2002. A review of the environmental management program. Presented to the Assistant Secretary for Environmental Management by the Top-to-Bottom Review Team, Washington, D.C., February 4.

Duffey, J. 2001. Testing hydrogen getters for TRUPACT-II payload expansion. Presented to the National Research Council's Committee on Long-Term Research Needs for Managing Transuranic and Mixed Wastes at Department of Energy Sites. Aiken, S.C., September 20.

Eaton, D., and T. Carlson. 2002. TMFA regulatory assistance. Presented to the TRU and Mixed Waste Focus Area End-User Review, Baltimore, Md., March 19-21.

Elias, D.A., L.R. Krumholz, R.S. Tanner, and J.M. Suflita. 1999. Estimation of methanogen biomass by quantitation of coenzyme M. Appl. Environ. Microbiol. 65(12):5541-5545.

EPA (U.S. Environmental Protection Agency). 2002a. Mixed Waste Glossary. EPA Radiation Protection Program Waste Management Team. Available at: <http://www.epa.gov/radiation/mixed-waste/mw_pg5.htm>.

EPA. 2002b. Cleanup Tools: Behind the Scenes at a Superfund Site. Washington, D.C.: U.S. Environmental Protection Agency. Available at: <http://www.epa.gov/oerrpage/superfund/accomp/400/toos.htm>.

Fortin, N.Y., and M.A. Deshusses. 1999. Treatment of methyl *tert*-butyl ether vapors in biotrickling filters. 1. Reactor startup, steady-state performance, and culture characteristics. Environmental Science and Technology 33(17):2980-2986.

Frazee, C., and S. Lott. 2002. Material Handling. Presentation to the TRU and Mixed Waste Focus Area End-User Review, Baltimore Md., March 19-21. Available at: <http://tmfa.inel.gov/Documents/MY02Unique.pdf>.

GAO (U.S. General Accounting Office). 1997. Department of Energy's Pit 9 Cleanup Project Is Experiencing Problems. GAO/T-RCED-97-221. Washington, D.C.: GAO. July.

GAO. 2001. Land Mine Detection: DOD's Research Program Needs a Comprehensive Evaluation Strategy. GAO-01-239. Washington, D.C.: GAO. April. Available at: <http://www.pnl.gov/nsd/commercial/tnd/Images/d01239.pdf>.

Harkness, M.R. 2000. Economic considerations in enhanced anaerobic biodegradation. Pp. 9-14 in Bioremediation and Phytoremediation of Chlorinated and Recalcitrant Compounds. G.B. Wickramanayake, A.R. Gavaskar, B.C. Alleman, and V.S. Magar, eds. Columbus, Ohio: Battelle Press.

Haschke, J.M., T.H. Allen, and L.A. Morales. 2000. Reaction of plutonium dioxide with water: Formation and properties of PuO_{2+x}. Science 287(Jan. 14):285-287.

Heitkamp, M. 2001. Recent scientific advancements have redefined the potential for bioprocesses. Paper presented to the National Research Council's Committee on Long-Term Research Needs for Managing Transuranic and Mixed Wastes at Department of Energy Sites. Aiken, S.C. September 20.

Ho, C.K., M.T. Itamura, M. Kelley, and R.C. Hughes. 2001. Review of Chemical Sensors for In-Situ Monitoring of Volatile Contaminants. SAND2001-0643. Albuquerque, N.M.: Sandia National Laboratories. Available at: <http://www.sandia.gov/sensor/SAND2001-0643.pdf>.

Hulet, G. 2002. Unique waste solutions. Presented to the TRU and Mixed Waste Focus Area End-User Review, Baltimore Md., March 19-21. Available at: <http://tmfa.inel.gov/Documents/MY02Unique.pdf>.

IAEA (International Atomic Energy Agency). 1993. Radioactive Waste Management Glossary. Vienna: IAEA.

Ji, H.F., E. Finot, R. Dabestani, T. Thundat, G.M. Brown, and P.F. Britt. 2000. A novel self-assembled monolayer coated microcantilever for low level caesium detection. Chemical Communications 6:457-458.

Ji, H.F., T. Thundat, R. Dabestani, G.M. Brown, B.F. Britt, and P.V. Bonnesen. 2001. Ultrasensitive detection of CrO_4^{2-} using a micro-cantilever sensor. Analytical Chemistry 73(7):1572-1576.

John, S.G., C.E. Ruggiero, L.E. Hersman, C-S Tung, and M.P. Neu. 2001. Siderophore mediated plutonium accumulation by *Microbacterium flavescens* (JG-9). Environmental Science and Technology 35(14):2942-2948.

Lange, C.C., L.P. Wackett, K.W. Minton, and M.J. Daly, 1998. Engineering a recombinant *Deinococcus radiodurans* for organo-pollutant degradation in radioactive mixed waste environments. Nature Biotechnology 16:929-933.

Leary, K.D. 2002. Research and development for a remotely operated, multi-purpose robotic vehicle. Paper presented to the National Research Council's Committee on Long-Term Research Needs for Managing Transuranic and Mixed Wastes at Department of Energy Sites. Richland, Wash. February 12.

Maio, V., and S. Reese. 2002. A survey of commercial MLLW treatment services and disposal options. Presented to the TRU and Mixed Waste Focus Area End-User Review, Baltimore, Md., March 19-21. Available at: <http://tmfa.inel.gov/Documents/MY02Vendor.pdf>.

Methvin, R. 2002. Excavation and Remediation of the Sandia National Laboratories Chemical Waste Landfill: Challenges and Successes. Abstract of a Master's Project. Department of Civil Engineering, University of New Mexico, Albuquerque, N.M., May.

Moody, D. 2002. TRU optimization thrust. Presented to the TRU and Mixed Waste Focus Area End-User Review, Baltimore, Md, March 19-21. Available at: <http://tmfa.inel.gov/Documents/MY02TRUOPT.pdf>.

Morris, M.I., I. Osborne-Lee, and G.A. Hulet. 2002. Demonstration of New Technologies Required for the Treatment of Mixed Waste Contaminated with > 260 ppm Mercury. ORNL/TM-2000/147. Oak Ridge, Tenn.

Muyzer, G. 1999. DGGE/TGGE a method for identifying genes from natural ecosystems. Current Opinion in Microbiology 2(3):317-322.

Nelson, H.H., and J.R. McDonald. 2001. Multisensor towed array detection system for UXO detection. IEEE Transactions on Geoscience and Remote Sensing 39:1139-1145.

Newman, D.K., and J.F. Banfield. 2002. Geomicrobiology: How molecular-scale interactions underpin biogeochemical systems. Science. 296:1071-1077.

Nichols, D.S., and T.A. McMeekin. 2002. Biomarker techniques to screen for bacteria that produce polyunsaturated fatty acids. Journal of Microbiological Methods 48(2-3):161-170.

NRC (National Research Council). 1995. Alternatives for Ground Water Cleanup. Washington, D.C.: National Academy Press.

NRC. 1996. The Waste Isolation Pilot Plant: A Potential Solution for the Disposal of Transuranic Waste. Washington, D.C.: National Academy Press.

NRC. 1997a. Building an Effective Environmental Management Science Program: Final Assessment. Washington, D.C.: National Academy Press.

NRC. 1997b. Innovations in Ground Water and Soil Cleanup: From Concept to Commercialization. Washington, D.C.: National Academy Press.

NRC. 1999a. The State of Development of Waste Forms for Mixed Wastes: U.S. Department of Energy's Office of Environmental Management. Washington, D.C.: National Academy Press.

NRC. 1999b. Technologies for Environmental Management: The Department of Energy's Office of Science and Technology. Washington, D.C.: National Academy Press.

NRC. 2000a. Research Needs in Subsurface Science: U.S. Department of Energy's Environmental Science Program. Washington, D.C.: National Academy Press.

NRC. 2000b. Alternatives for High-Level Waste Salt Processing at the Savannah River Site. Washington, D.C.: National Academy Press.

NRC. 2000c. Long-Term Institutional Management of U.S. Department of Energy Legacy Waste Sites. Washington, D.C.: National Academy Press.

NRC. 2001a. Research Needs for High-Level Waste Stored in Tanks and Bins at U.S. Department of Energy Sites: Environmental Management Science Program. Washington, D.C.: National Academy Press.

NRC. 2001b. Research Opportunities for Deactivating and Decommissioning Department of Energy Facilities. Washington, D.C.: National Academy Press.

NRC. 2001c. Disposition of High-Level Waste and Spent Nuclear Fuel: The Continuing Societal and Technical Challenges. Washington, D.C.: National Academy Press.

NRC. 2001d. Improving Operations and Long-Term Safety of the Waste Isolation Pilot Plant. Final Report. Washington, D.C.: National Academy Press.

NRC. 2001e. A Strategic Vision for the Department of Energy Environmental Quality Research and Development. Washington, D.C.: National Academy Press.

NRC. 2002a (forthcoming). Improving the Scientific Basis for Managing Nuclear Materials and Spent Nuclear Fuel Through the Environmental Management Science Program. Washington, D.C.: National Academy Press.

NRC. 2002b. Characterization of Remote-Handled Transuranic Waste for the Waste Isolation Pilot Plant. Final report. Washington, D.C.: National Academy Press.

Owendoff, J. 2002. New Directions in Science and Technology for Environmental Management: Action Plan. Presented to the National Research Council's Board on Radioactive Waste Management. Washington, D.C., April 16.

Perkins, B.L. 1976. Incineration Facilities for Treatment of Radioactive Wastes: A Review. Los Alamos Scientific Laboratory Report UC-70. July. Washington, D.C.: U.S. Energy Research and Development Administration.

Pierce, B. 2001. Pu-238 Decontamination. Presented to the National Research Council's Committee on Long-Term Research Needs for Managing Transuranic and Mixed Wastes at Department of Energy Sites. Aiken, S.C., September 20.

Pottie, G.J., and W.J. Kaiser. 2000. Wireless Integrated Network Sensors. Communications of the ACM (Association of Computing Machinery) 43(5):51-58.

Reed, D.T., Y. Vojta, J.W. Quinn, and M.K. Richmann, 1999. Radiotoxicity of plutonium in NTA-degrading *Chelatobacter heintzii* cell suspensions. Biodegradation 10(4):251-260.

Reysenbach, A.-L., and E. Shock. 2002. Merging genomes with geo-chemistry in hydrothermal ecosystems. Science 296:1077-1082.

Rondon, M.R., P.R. August, A.D. Bettermann, S.F. Brady, T.H. Grossman, M.R. Liles, K.A. Loiacono, B.A. Lynch, I.A. MacNeil, C. Minor, C.L. Tiong, M. Gilman, M.S. Osburne, J. Clardy, J. Handelsman, and R.M. Goodman. 2000. Cloning the soil metagenome: A strategy for accessing the genetic and functional diversity of uncultured microorganisms. Applied and Environmental Microbiology 66(6):2541-2547.

Rusin, P.A., L. Quintana, J.R. Brainard, B.A. Strietelmeier, C.D. Tait, S.A. Ekberg, P.D. Palmer, T.W. Newton, and D.L. Clark, 1994. Solubilization of plutonium hydrous oxide by iron-reducing bacteria. Environmental Science and Technology 28(9):1686-1690.

Rütter, H., H. Sass, H. Cypionka, and J. Rullkötter, 2002. Phospholipid analysis as a tool to study complex microbial communities in marine sediments. Journal of Microbiological Methods 48(2-3):149-160.

St. Michel, W., and S. Lott. 2002. Non-destructive Characterization of MLL and MTRU Waste. Presented to the TRU and Mixed Waste Focus Area End-User Review, Baltimore, Md., March 19-21. Available at: <http://tmfa.inel.gov/Documents/MY02Char.pdf>.

Sandia National Laboratories (SNL). 1998. Robotics and Intelligent Machines in the U.S. Department of Energy. A Critical Technology Roadmap. SAND98-2401/2. October. Available at: <http://www.sandia.gov/isrc/RIMfinal.pdf>.

Schwinkendorf, W.E., B.C. Musgrave, and R.N. Drake. 1997. Evaluation of Alternative Nonflame Technologies for Destruction of Hazardous Organic Waste. INEL/EXT-97-00123. April. Idaho Falls, Idaho: Idaho National Engineering and Environmental Laboratory. Available at: <http://tmfa.inel.gov/Documents/nonflame.pdf>.

Steffan, R.J., S. Vainberg, C.W. Condee, K. McClay, and P. Hatzinger. 2000. Biotreatment of MTBE with a new bacterial isolate. Pp. 165-173 in Bioremediation and Phytoremediation of Chlorinated and Recalcitrant Compounds (C2-4), G.B. Wickramanayake, A. R. Gavaskar, B.C. Alleman, and V. S. Magar, eds. Columbus, Ohio: Battelle Press.

Stemmer, W.P.C. 1994. Rapid evolution of a protein in vitro by DNA shuffling. Nature 370(6488):389-391.

Takai, K., D.P. Moser, M. DeFlaun, T.C. Onstott, and J. K. Fredrickson. 2001. Archaeal diversity in waters from deep South African gold mines. Appl. Environ. Microbiol. 67:5750-5760.

Townsend, J. 2001. The TRU and Mixed Waste Focus Area—Bridging the gap between waste inventory and disposal methodologies. Radwaste Solutions. 8(1):20-24.

Triay, I. 2001. National TRU waste program vision for the future. Presented to the National Research Council's Committee on Long-Term Research Needs for Managing Transuranic and Mixed Wastes at Department of Energy Sites. Carlsbad, N.M., December 5.

Unterman, R., M.F. DeFlaun, and R.J. Steffan. 2000. Advanced in situ bioremediation—A hierarchy of technology choices. Pp. 399-414 in Biotechnology, Vol. 11b, Environmental Processes II, J. Klein, ed. New York: Wiley-VCH.

Wagner-Döbler, I., H. von Canstein, Y. Li, K.N. Timmis, and W-D Deckwer. 2000. Removal of mercury from chemical wastewater by microorganisms in technical scale. Environmental Science and Technology 34(21):4628-4634.

Wen, S., and D.D.L. Chung. 2001. Rectifying and thermocouple junctions based on Portland cement. Journal of Materials Research 16(7):1989-1993.

White, O., J.A. Eisen, J.F. Heidelberg, E.K. Hickey, J.D. Peterson, R.J. Dodson, D.H. Haft, M.L. Gwinn, W.C. Nelson, D.L. Richardson, K.S. Moffat, H. Qin, L. Jiang, W. Pamphile, M. Crosby, M. Shen, J.J. Vamathevan, P. Lam, L. McDonald, T. Utterback. C. Zalewski, K.S. Makarova, L. Aravind, M.J. Daly, K.W. Minton, R.D. Fleischmann, K.A. Ketchum, K.E. Nelson, S. Salzberg, H.O. Smith, J.C. Venter, and C.M. Fraser. 1999. Genome sequence of the radioresistant bacterium *Deinococcus radiodurans* R1. Science 286(Nov. 19):1571-1577.

Wildung, R.E., and T.R. Garland. 1982. Effects of plutonium on soil microorganisms. Applied and Environmental Microbiology 43:418-423.

Zhang, C.L. 2002. Stable carbon isotopes of lipid biomarkers: Analysis of metabolites and metabolic fates of environmental microorganisms. Current Opinion in Biotechnology 13(1):25-30.

Appendixes

A

Overview of the Environmental Management Science Program and Pending Changes

The Department of Energy (DOE's) site cleanup program is one of the largest environmental cleanup efforts in world history. The program is currently estimated to cost more than $220 billion (DOE, 2002). To deal with this task, DOE established its Office of Environmental Management (EM) in November 1989. As this report was being finished, EM completed a "top-to-bottom" review of its programs (DOE, 2002). The review led to significant changes, which will be finalized at the start of fiscal year 2003 pending approval by Congress. This appendix presents an overview of the EM Office of Science and Technology (OST) and the Environmental Management Science Program (EMSP) before the top-to-bottom review. It also describes elements of the restructured OST and EMSP known to the committee in spring 2002.

The EM Office of Science and Technology

The Office of Science and Technology is the EM office charged with developing new technologies to assist the cleanup mission. Research and development investments by OST have the following objectives:

- To meet the high-priority needs identified by the cleanup project managers.
- To reduce the cost of EM's costliest cleanup projects.
- To reduce the technological and programmatic risk of completing major cleanup projects on time and within budget.
- To accelerate and increase technology deployments (DOE, 2000e).

OST used three main approaches to achieving its objectives: (1) site technology coordinating groups (STCGs), (2) focus areas, and (3) the EM Science Program.

To identify science and technology needs, OST formed an STCG at each major cleanup site to interact with local contractor personnel and others directly involved in the cleanup activities. Each group included a senior manager from the site DOE office, site contractors, and national laboratory personnel. The STCGs were responsible for developing and prioritizing a list of site problems and technology needs based on environmental management issues relevant to a specific site.

Beginning in 1995 and continuing through most of the committee's study period for this report, OST's activities were organized around five focus areas:

1. deactivation and decommissioning
2. high-level waste tanks
3. subsurface contaminants
4. transuranic and mixed waste
5. nuclear materials

The primary role of the focus areas was to identify, develop, and deploy new technologies to meet site needs. The National Research Council (NRC) provided a number of studies and reports for the focus areas, as well as a summary report (NRC, 1999b).

The Environmental Management Science Program

The EMSP was established in response to a mandate from Congress in the fiscal year 1996 Energy and Water Development Appropriations Act. Congress directed DOE to "provide sufficient attention and resources to longer-term basic science research which needs to be done to ultimately reduce cleanup costs, . . . develop a program that takes advantage of laboratory and university expertise, and . . . seek new and innovative cleanup methods to replace current conventional approaches which are often costly and ineffective" (DOE, 2000a, p.1). Research supported by the EMSP is expected to lead to new knowledge and technologies that reduce the costs, schedule, and risks associated with the most challenging technical problems in DOE's site cleanup program. From its inception in 1996 through fiscal year 2001, the EMSP has provided $294 million in

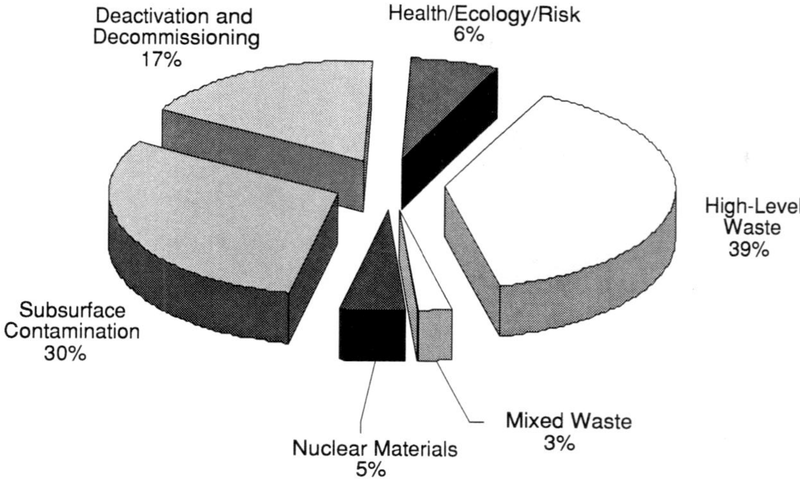

FIGURE A.1 Breakdown of EMSP funding by problem area in fiscal year 2002. The EMSP budget request for fiscal year 2003 is about $30 million. Source: DOE Environmental Management Science Program.

funding for 361 research projects. Information on how the EMSP develops its requests for research proposals, evaluates proposals, and funds research was presented in three previous reports (NRC 2000a, 20001a, 2001b).

The DOE Office of Environmental Management and the DOE Office of Science jointly managed the EMSP.[1] EM's OST had lead responsibility, and the EMSP was closely tied to OST's focus areas. Each year, the EMSP issued calls for research proposals related to one or two of OST's focus areas—this report was prepared to assist in a planned call for proposals in the transuranic and mixed waste area in fiscal year 2003. Research proposals and their funding were tracked according to the focus areas, as indicated in Figure A.1 and Table A.1. Current planning is to move the EMSP out of EM entirely and into the Office of Science under its Office of Biological and Environmental Research (OBER).[2]

A two-part review of research proposals—for scientific merit by the Office of Science and for relevance to site cleanup by EM—has been a key feature of the EMSP. The relevance review was based largely on site needs identified by the STCGs and the focus areas,

[1]EM and the Office of Science are two of eight DOE offices that report to the Under Secretary for Energy, Science, and Environment. The Office of Science is responsible for most of DOE's basic scientific research.

[2]Environmental remediation sciences in OBER will include the Environmental Molecular Sciences Laboratory at the Pacific Northwest National Laboratory, the Natural and Accelerated Bioremediation Research program, the Savannah River Ecology Laboratory, and the EMSP.

TABLE A.1 EMSP Projects in Transuranic and Mixed Wastes

Project Title	Funding
Removal of Heavy Metals and Organic Contaminants from Aqueous Streams by Novel Filtration Methods	$330,000
Architectural Design Criteria for F-Block Metal Ion Sequestering Agents	$1,800,000
A Novel Energy-Efficient Plasma Chemical Process for the Destruction of Volatile Toxic Compounds	$980,222
Extraction and Recovery of Mercury and Lead from Aqueous Waste Streams Using Redox-Active Layered Metal Chalcogenides	$333,000
Utilization of Kinetic Isotope Effects for the Concentration of Tritium	$1,354,000
An Alternative Host Matrix Based on Iron Phosphate Glasses for the Vitrification of Specialized Nuclear Waste Forms	$624,834
Acid-Base Behavior in Hydrothermal Processing of Wastes	$379,620
High Fluence Neutron Source for Nondestructive Characterization of Nuclear Waste	$745,139
New Anion-Exchange Resins for Improved Separations of Nuclear Materials	$1,212,211
Managing Tight-Binding Receptors for New Separations Technologies	$350,000
Processing of High-Level Waste: Spectroscopic Characterization of Redox Reactions in Supercritical Water	$112,000
Photocatalytic and Chemical Oxidation of Organic Compounds in Supercritical Carbon Dioxide	$660,000
Supramolecular Chemistry of Selective Anion Recognition for Anions of Environmental Relevance	$775,000
The Sonophysics and Sonochemistry of Liquid Waste Quantification and Remediation	$769,843
Spectroscopy, Modeling, and Computation of Metal Chelate Solubility in Supercritical CO_2	$265,937
The Adsorption and Reaction of Halogenated Volatile Organic Compounds (VOCs) on Metal Oxides	$390,000
Adsorption/Membrane Filtration as a Contaminant Concentration and Separation Process for Mixed Wastes and Tank Wastes	$609,987
Development of Advanced In Situ Techniques for Chemistry Monitoring and Corrosion Mitigation in Supercritical Water Oxidation Environments	$696,395
De Novo Design of Ligands for Metal Separation	$380,000
Ion and Molecule Sensors Using Molecular Recognition in Luminescent, Conductive Polymers	$1,500,000
Fundamental Chemistry and Thermodynamics of Hydrothermal Oxidation Processes	$1,220,000
Photooxidation of Organic Waste Using Semiconductor Nanoclusters	$1,251,000
Hazardous Gas Production by Alpha Particles in Solid Organic Transuranic Waste Matrices	$400,362
Real-Time Broad Spectrum Characterization of Hazardous Waste by Membrane Introduction Mass Spectrometry	$655,000
The Development of Cavity Ringdown Spectroscopy as a Sensitive Continuous Emission Monitor for Metals	$538,000
Rational Synthesis of Imprinted Organofunctional Sol-Gel Materials for Toxic Metal Separation	$450,000
Genetic Engineering of a Radiation-Resistant Bacterium for Biodegradation of Mixed Wastes	$442,398
Measurements and Models for Hazardous Chemical and Mixed Wastes	$500,000
Novel Miniature Spectrometer for Remote Chemical Detection	$549,000
Isolation of Metals from Liquid Wastes: Reactive Scavenging in Turbulent Thermal Reactors	$1,075,000
Rational Design of Metal Ion Sequestering Agents	$405,000
Genetic Engineering of a Radiation-resistant Bacterium for Biodegradation of Mixed Wastes	$480,000
Miniature Chemical Sensor Combining Molecular Recognition with Evanescent-Wave Cavity Ring-Down Spectroscopy	$949,999
Radiation Effects on Materials in the Near-Field of Nuclear Waste Repository	$450,000

along with input from advisory groups such as this and previous NRC committees (NRC 2000a, 2001a, 2001b). Without the focus area structure it may be more difficult for the EMSP to identify site technology needs and especially to keep a perspective on long-term needs that can be addressed through scientific research. Maintaining the relevance of its funded research to site cleanup needs will be important for the EMSP after the restructuring is completed in fiscal year 2003—for example, by continuing the joint review of EMSP proposal by both OST for relevance to EM's needs and the Office of Science for scientific merit.

The Reoganization of OST

In an effort to expedite and reduce the ultimate costs of DOE site cleanup, the Secretary of Energy directed that a review of the EM program be undertaken. In August the Assistant Secretary for EM created a Top-to-Bottom Review Team, which issued its findings in February 2002 (DOE, 2002). As a result of the review, OST began a reorganization that is intended to, inter alia,

- optimize and fast-track the use of science and technology for EM cleanup projects;
- concentrate on high-risk, high-cost problems; and
- focus only on activities that promise high payback or step improvements.

In place of the focus areas, OST is to be structured around two "thrusts" that are identified as closure site support and alternatives and step improvements to current high-risk, high-cost baselines (see Sidebar A.1).

SIDEBAR A.1 NEW THRUSTS FOR THE EM OFFICE OF SCIENCE AND TECHNOLOGY

OST's new thrusts, which are to replace the focus areas, were presented to the NRC Board on Radioactive Waste Management in April 2002. These thrusts, objectives, and strategies as developed at that time were the following:

Thrust 1: Closure Site Support

Objectives:

- **Ensure that the closure sites (Rocky Flats and Ohio) have the necessary technology and technical support to meet closure schedules.**
- **Provide science and technology to improve baselines and schedules at all small sites.**

Strategies:

- **Establish a multidisciplinary, hands-on technical team dedicated to assisting the closure sites.**
- **Provide a dedicated budget to ensure that necessary resources are readily available.**
- **Streamline the science and technology proposal process to ensure real-time response to needs, while ensuring high-quality work.**

Thrust 2: Alternative and Step Improvements to Current High-Risk, High-Cost Baselines

Objectives:

- **Ensure that cleanup goals can be accomplished at reasonable costs and schedules.**
- **Ensure that all possible cleanup alternatives are evaluated.**
- **Ensure that improved, workable alternatives are available and utilized as the cleanup progresses.**

Strategies:

- **Establish a focused, headquarters-directed science and technology program to address alternatives to current plans.**
- **Identify technology areas where the greatest benefit could be realized by an aggressive investment strategy.**

SOURCE: Owendoff, 2002.

B

The Transuranic and Mixed Waste Inventory

The accumulation of radioactive wastes began in the 1940s with the development of the atomic bomb and continued with the large-scale production of fissile materials such as uranium and plutonium during the Cold War period. Manufacturing processes involved the production of plutonium and its separation from irradiated fuel elements, the development and application of methods for isotopic enrichment, and the production and fabrication (casting, machining, plating) of metal at Hanford, Washington; Rocky Flats, Colorado; Oak Ridge, Tennessee, and other supporting sites. The lower-activity wastes from these operations ranged from trash contaminated with plutonium, to process wastes (e.g., organic sludges or waste contaminated with metallic compounds) from liquid-liquid extraction employed in product purification. During this period, emphasis was placed on production and little attention was given to the types or quantities of waste generated. Wastes were managed using practices analogous to those found in other process industries, which involved the use of on-site disposal in landfills for process waste and the use of ponds and lagoons to control large volumes of wastewater. Wastes were frequently contaminated with both radioactive and chemical substances.

Transuranic and Mixed (TM) Wastes

The radioactive wastes from these processes have been categorized into two types, transuranic waste (TRU)[1] and low-level waste (LLW).[2]

---

[1]TRU is radioactive waste that contains more than 100 nanocuries of alpha-emitting isotopes per gram of waste, with atomic numbers greater than 92 and half-lives greater than 20 years (DOE Order 435.1, issued July 1999). This waste results primarily from fuel reprocessing and from the fabrication of plutonium weapons and plutonium-bearing reactor fuel. Generally, little or no shielding is

Mixtures of TRU or LLW with toxic or hazardous substances, defined by the Resource Conservation and Recovery Act (RCRA) as well as the Toxic Substances Control Act (TSCA) and any applicable state regulations, are defined as mixed wastes and identified as MTRU or MLLW.

Exposure to radioactivity was recognized as a human health hazard at the onset of nuclear operations, and standards for health protection were established early on by the Atomic Energy Commission, forerunner to the Department of Energy (DOE) and the U.S. Nuclear Regulatory Commission (USNRC). These standards constrained the manner in which radioactive materials were handled to minimize human exposure. In 1970, new standards were established for the burial of transuranic wastes.

Human health hazards resulting from exposure to chemicals were not fully recognized until 1970, with the formation of the Environmental Protection Agency (EPA) and the subsequent establishment of air and water pollution standards. Substantial control of chemical waste disposal began in 1976 with the authorization of RCRA, which directed EPA to set standards for the land disposal of toxic and hazardous substances. Prior to this time, little attention was given to controlling chemical wastes. Compounds of lead, mercury, cadmium, and chromium along with commonly used industrial solvents were allowed to enter the environment with little control.

The application of USNRC and EPA regulations has further subdivided the waste. TRU wastes generated after 1970 have been placed in retrievable storage. In response to RCRA regulations, ponds and lagoons were closed and chemically contaminated trash was no longer buried but instead placed in retrievable storage. The waste materials have several attributes summarized below and discussed further in succeeding paragraphs:

- The volume of waste is substantial.
- The waste is incompletely characterized as to its physical state and its radiological and chemical components.
- The waste is highly heterogeneous on both the total volume and the individual waste container scale.
- The chemical or radiological contents of many waste streams contain individual substances or mixtures of components that complicate the selection and application of treatment systems.

---

required, but energetic gamma and neutron emissions from certain TRU nuclides and fission product contaminants may require shielding or remote handling.

[2]LLW is any radioactive waste including accelerator-produced waste that is not classified as spent nuclear fuel, high-level waste, TRU waste, 11e(2) by-product material, or naturally occurring radioactive material (DOE Order 435.1, issued July 1999).

Current and Expected Waste Volumes

Present estimates of the volumes of waste represented by the categories of TRU and MLLW are found in the DOE's Summary Data on the Radioactive Waste, Spent Nuclear Fuel, and Contaminated Media Managed by the U.S. Department of Energy (DOE, 2001a). Table B.1, which is excerpted from this document, illustrates the magnitude of the TRU and MLLW waste volume reported at the end of fiscal year 1999 or 2000. DOE does not distinguish between TRU and MTRU in the summary, essentially because all of DOE's TRU, including MTRU, is destined for disposal in the Waste Isolation Pilot Plant (WIPP; see Chapter 2).

The chemical composition of the previously buried waste is largely unknown. Because production processes did not change with the 1970 limitation on burial, it probably has similar composition to waste in retrievable storage. The fate of previously buried waste is yet to be determined, but it is expected that some of this material will require retrieval.

The summary contains no information about the physical or chemical characteristics of the waste materials. Information about radioactivity is reflected only by the materials' classification as TRU or LLW. Other sources of information are necessary to complete the description of the inventory.

For the purpose of developing site treatment plans in the early 1990s, the Department of Energy directed all sites to evaluate the inventory of accumulated TRU and mixed waste based on the best information

TABLE B.1 Overview of the DOE's Transuranic and Mixed Wastes

	Volume	
Origin	TRU (m^3)	MLLW (m^3)
Buried (pre-1970)	137,000	317,000[a]
Retrievably stored (1970-1999)	111,000	44,500
Predicted new waste generation	60,000[b]	100,000[c]
Recovered soils and sediments (2002-2010)	32,000	170,000

[a] α-LLW.
[b] 2000-2034.
[c] 2000-2070.

SOURCE: DOE, 2001a.

available (DOE, 1997). This inventory was very detailed and based on the best available knowledge, including available sampling and analysis, history of the process that generated the waste, or the recollections of persons involved in manufacturing operations. Information was gathered for each waste stream, including radioactive materials, chemical constituents, a text description of the waste, and a physical description of the matrix (e.g., solid, liquid, debris). Because of the uncertainty in knowledge, a degree of confidence (high, medium, or low) was also assigned to the data established for each waste stream in the categories of matrix or physical state, chemical composition, and radioactivity. Although the inventory lacks high confidence in characterization of all waste streams, it does allow study of the potential problems of treatment and disposal confronting DOE. The following discussion and tables are based on information in this database and apply to wastes that were inventoried in 1995.

Diversity of the Inventory

Information about the wastes is crucial to the design of treatment processes and the generation of data supporting disposal requirements. Table B.2 illustrates overall knowledge about the waste as expressed by the confidence in the data. There is reasonable confidence in the chemical composition for only about a third of the MLLW and pond residue waste volume. The chemical characterization of the MTRU waste is even poorer, with only a sixth of the volume meeting the criteria of medium to high confidence. In addition to the need for better data to support treatment and disposal requirements, some reclassification of MTRU and MLLW wastes is expected, as the composition is better determined.

The cover of this report shows cross sections of waste drums as examples of the typical heterogeneity of the wastes. Sidebar B.1, taken from a previous National Research Council (NRC) study, illustrates the

TABLE B.2 Percentage of Inventoried Waste Described with a Medium or High Degree of Confidence (by volume)

Category	TRU	MTRU	MLLW	Pond Residue
Matrix or physical state	58	91	65	99
Chemical composition	N/A	16	35	34
Radioactivity	96	99	48	54

Note: N/A = not available.

diversity of materials that are found in mixed waste. Table B.3 shows the relative magnitude of retrievably stored and pond residue wastes in each of the five categories described in the sidebar.

The retrievably stored MLLW, the pond residues, and the previously buried MLLW or pond waste that is excavated must be treated to meet RCRA Land Disposal Restrictions and the requirements of the USNRC for disposal of radioactive materials. Chemical and radiological composition becomes a significant issue in the selection, design, and operation of treatment systems.

The retrievably stored TRU and MTRU waste and any previously buried TRU waste that is recovered are destined for disposal at the WIPP site and must meet the special requirements for shipment and acceptance at that site. Treatment to meet RCRA Land Disposal Restrictions is not required. However the TRU and MTRU waste must meet the requirements of TSCA for the treatment of polychlorinated biphenyls (PCBs).

MLLW and Pond Residue Waste

These wastes contain chemicals designated as hazardous by the Environmental Protection Agency under the Resource Conservation and Recovery Act as well as low levels of radioactive fission products.

Chemical Composition

Table B.4 identifies the fractions of the waste volume in each category known or suspected to be contaminated with various classes of hazardous and toxic materials. These classes are chosen to reflect chemical contaminants commonly found in mixed waste and represent major processes used for the treatment and separation of chemical wastes prior to disposal in a RCRA facility.

- *Mercury:* Mercury occurs in several forms including the metal, amalgams with other metals, inorganic compounds, and organic compounds. Each form of mercury requires a different approach to treatment. Because of its high vapor pressure, mercury poses a problem when exposed to high temperatures such as those encountered in incineration where it vaporizes and enters the off-gas, from which it is difficult to trap and remove. Grouting techniques useful for other heavy metals are generally not effective to control mercury and its compounds. Retorting is used in the chemical industry for the removal of mercury from waste and to prepare it for recycling. The applicability of this technique to separate mercury from radioactive materials is unknown.

SIDEBAR B.1 A PERSPECTIVE ON TM WASTE DIVERSITY

For the purpose of developing site treatment plans for TM wastes, DOE established five treatment groups. The types of waste included in each group provide a perspective on the overall waste diversity.

Debris

- **It contains one or more of over 700 materials listed as hazardous by the EPA;**
- ***Metal Debris:* Metal with or without lead or cadmium**
- ***Inorganic Nonmetal Debris:* Concrete, glass, ceramic or brick, rock, asbestos, and graphite**
- ***Organic Debris:* Plastic or rubber, leaded gloves or aprons, halogenated plastics, nonhalogenated plastics, wood, paper, and biological matter**
- ***Heterogeneous Debris:* Composite filters, asphalt, electronic equipment, and other inorganic and organic materials**

Inorganic Homogeneous Solids and Soils

- ***Inorganic Homogeneous Solids:* Particulate matter—such as ash, sandblasting media, inorganic particulate absorbents, absorbed organic liquids, ion-exchange media, metal chips or turnings, glass or ceramic materials, and activated carbon**
- ***Inorganic Sludges:* Wastewater treatment pond, off-gas treatment, plating waste, and low-level reprocessing sludges**
- ***Other Inorganic Waste:* Paint waste (chips, solids, and sludges), salt waste containing chlorides, sulfates, nitrates, metal oxides or hydroxides, and inorganic chemicals**
- ***Solidified Homogeneous Solids:* Soil, soil/debris, and rock/gravel**

TABLE B.3 Percentage Distribution of Inventoried Waste by Volume

Category	TRU	MTRU	MLLW	Pond Residue
Aqueous		0.1	9	
Organic	2	0.5	4	
Solids	2	28	25	100
Debris	95	71	57	
Unique	1	0.1	4	

Organics

- ***Organic Liquids:*** **Aqueous streams containing both halogenated and nonhalogenated organic compounds as well as pure organic streams containing halogenated and nonhalogenated compounds**
- ***Organic Homogeneous Solids:*** **Organic particulate matter (resins, organic absorbents), organic sludges (biological, halogenated, and nonhalogenated), and organic chemicals**

Unique Waste

- ***Lab Packs:*** **Organic, aqueous, and solid laboratory chemicals and scintillation cocktails**
- ***Special Wastes:*** **Elemental mercury, elemental hazardous metals (activated and nonactivated lead, elemental cadmium), beryllium dust, batteries (lead acid, mercury, cadmium), reactive metals (bulk and reactive metal-contaminated components), pyrophoric fines, explosives, or propellants, and compressed gases and aerosols**
- ***All Others:*** **Materials placed in a final waste form are included in this category**

Wastewaters

- **Acidic, basic, and neutral aqueous liquids and slurries, including cyanide-containing wastewaters and slurries**

SOURCE: DOE, 1995.

- *Metallic compounds (metals):* Compounds containing elements such as chromium, cadmium, and lead are commonly treated by grouting, which converts these materials to an insoluble form, resistant to leaching and acceptable for land disposal.
- *Toxic organic materials and solvents (organics and solvents):* This category includes halogenated and nonhalogenated solvents commonly found in industry and declared by EPA to be either toxic or hazardous. Incineration is commonly employed to remove and destroy these materials. Encapsulation and grouting are generally not effective treatments for this type of contaminant.

TABLE B.4 Percentage of Inventoried Waste Volume Contaminated with RCRA Materials

	MLLW					
Classes	Aqueous	Organic	Solids	Debris	Unique	Pond Solids
Metals only	17	2	17	7	44	50
Metals with mercury	54	2	4	5	9	1
Organics, solvents, metals, mercury	15	31	27	16	8	17
Total Mercury	**70**	**34**	**31**	**20**	**17**	**17**
Organics and solvents, metals	11	43	43	42	6	23
Total Metals	**98**	**79**	**90**	**70**	**66**	**90**
Organics/Solvents only	2	15	5	19	10	9
Total Organics and Solvents	**27**	**90**	**75**	**77**	**23**	**49**
PCBs	8	48	15	3	5	15
Plating waste	10	3	34	5	0.4	30
Hazardous characteristics	97	65	57	31	58	18
Others (including no data)		2	2	12	3	

- *PCBs:* These compounds are highly resistant to natural degradation and are known to accumulate in the fatty tissue of animals and humans. TSCA requires that PCBs be destroyed to an acceptable level. Incineration is a common treatment method.
- *Waste from electroplating and metal treatment (plating waste):* In addition to metallic ions, these wastes commonly contain cyanides, which must be destroyed either chemically or by incineration.
- *Hazardous characteristics:* Materials designated as ignitable, corrosive, or reactive by RCRA regulations are unacceptable for land disposal without treatment. The unique waste category contains a high percentage of reactive materials such as sodium-potassium alloy, pyrophoric materials, explosives, and compressed gases that pose special treatment problems.

Metals, mercury, and organics or solvents are selected as major chemical classes in Table B.4. In addition to presenting total data, the table shows the breakdown of commingled waste in each class. The data represent only that one or more chemicals in the particular class exceed the level allowed for land disposal under RCRA Land Disposal Restrictions (LDRs). No information is available regarding quantitative levels of contamination. In some instances, a waste stream is classified as MLLW, but there is no specific information as to the contaminating material.

Examination of Table B.4 illustrates the difficulties inherent in the selection and application of treatment systems. There are few waste streams that might be considered "pure" (i.e., other metals without mercury or organics and solvents, organics and solvents without mercury). For the most part, waste streams are heterogeneous and contain some of everything. For simplicity, the following discussion considers only the debris category (57 percent of the total MLLW volume). However, the conclusions apply to all categories. It must also be stressed that the data in the table do not represent a quantitative measure of the amounts of contaminating material in the wastes, only that the materials are probably present at levels requiring treatment.

Incineration is a common method for destroying RCRA organic materials. Control of air emissions from incineration or other thermal treatment methods is a major consideration in the employment of this technology. As noted above, 77 percent of the debris category contains total organics and solvents above levels acceptable for land disposal. However, 16 percent of the debris volume contains both organic materials and mercury. The ability of a thermal treatment system to accept this fraction will depend on the quantitative amounts of mercury and the efficiency of the system for the removal of mercury from discharged gases.

Retorting is commonly used for the separation of mercury from waste materials. It is carried out at lower temperatures than incineration and usually involves indirect heating to avoid mixing combustion gases with mercury vapor. In the debris category, 20 percent of the volume is contaminated with enough mercury to require treatment. However, 80 percent of this quantity is co-contaminated with organics and solvents, which will evolve with the mercury and significantly complicate the separation of the materials.

Alkaline grouts are common for stabilizing heavy metals, but they are not generally effective for controlling significant quantities of accompanying mercury or organic materials. The effectiveness of grouting depends on the quantities of contaminating materials. Wastes containing only small amounts of mercury or organic materials, only marginally above the required treatment levels, might be grouted successfully. Referring again to the debris group, 70 percent of the volume in this category is contaminated with other metals requiring treatment. However, 90 percent of the metal-contaminated waste in this category is co-contaminated with organics and/or mercury.

Hence, no single treatment system is likely to meet the requirements of all waste streams containing similar materials. Quantitative measurement of the constituents of the waste streams is key to the selection and operation of treatment systems.

Radioactivity

Meeting the requirements for RCRA land disposal is only one treatment consideration. Another consideration involves the levels and types of radioactivity associated with the waste. Knowledge about its radioactivity is necessary to specify the design of treatment systems that may be operated and maintained safely. In addition, specific isotopes and radioactivity levels in the waste will govern the disposal of treated MLLW. The USNRC specifies three classes of radioactivity for low-level waste—A, B, and C—each containing higher levels of radioactivity (see Chapter 2, Table 2.4). The permitting of disposal facilities requires a performance assessment of proposed sites to ensure adequate containment of the radioactivity.

Table B.5 shows levels of radioactivity associated with the various classes of waste. Generally, wastes at the lower end of the spectrum, <10 nanocuries of alpha-emitting TRU materials per gram, may qualify as Class A. Those at the higher end of the spectrum, approaching 100 nanocuries of alpha-emitting TRU materials per gram, are most likely Class C. Wastes that contain >100 nanocuries of alpha-emitting isotopes per gram are classified as TRU wastes and cannot be disposed as MLLW.

Certain actinides such as Pu-238 have attributes that may affect the choice of treatment processes. Oxides of Pu-238 are known to form extremely finely divided particles that are dispersed easily. The high radioactivity of this actinide and the toxicity of plutonium place added emphasis on selecting treatment systems that can manage this material properly. The presence of Pu-238 is not a quantitative measure but only an indicator that it is present in the waste and should be considered. Quantitative measurements of the waste must be made to understand

TABLE B-5 Selected Radioactivity Parameters of Inventoried Waste (Volume percent)

	MLLW					
	Aqueous	Organic	Solids	Debris	Unique	Pond Residue Solids
α emitters < 10 nCi/g	51	60	27	13	64	88
α emitters 10 - 100 nCi/g	5	8	57	59	10	11
α emitters unknown	44	32	16	28	25	1
Pu-238	1	11	37	58	11	39
Organics and solvents with Pu-238	0.4	10	32	54	3	20
Mercury with Pu-238	0.2	6	19	9	0.8	0.1

the significance of this contaminant and its effect on treatment system operation.

Wastes in the MLLW aqueous, organic, and unique categories and in pond residue solids have relatively low levels of alpha activity. They are candidates for disposal after treatment at facilities having both a RCRA and a USNRC Class A permit, such as Envirocare in Utah. However, some combinations of chemical and radioactive isotopes will require highly specialized treatment schemes. DOE has estimated that 10 to 15 percent of the mixed waste will fall into this category (DOE, 2001a).

TRU and MTRU Waste

Although free from RCRA treatment requirements, TRU and MTRU waste must meet transportation and waste acceptance criteria for WIPP. Shipments received at WIPP must also be characterized for RCRA components even though treatment is not required. Considerations for shipment and acceptance at WIPP follow:

- The generation of hydrogen gas during shipment must be controlled. Hydrogen generation may result from radiolysis of organic materials (e.g., wood, paper, cloth, plastics, solvents), biological activity, or corrosion. In the event of a transportation accident, an explosion of accumulated hydrogen might result in the dispersal of both hazardous and radioactive materials.
- Free liquids exceeding 1 volume percent of the outside container are prohibited. Wastes classified as aqueous or organic liquids are not acceptable.
- Most of the material comprising the unique waste category is prohibited. This includes explosives and compressed gases as well as pyrophoric materials.
- Corrosive wastes (defined by RCRA) are prohibited. However, corrosive waste usually can be treated easily.
- Flammable volatile organic compounds (VOCs) are limited to 500 parts per million (ppm) in the headspace of any payload container. PCBs are currently prohibited from disposal by the TSCA requirement that specifies destruction of this material. An effort is currently in progress to obtain administrative relief and allow disposal of PCBs at WIPP without treatment.
- Highly radioactive materials designated as remote-handled TRU (RH-TRU) are currently prohibited from disposal at WIPP. Work is in progress to define methods for safe shipment and handling of RH-TRU.

Some properties of TRU and MTRU wastes are listed in Table B.6.

TABLE B.6 Selected Properties of Inventoried MTRU and TRU Waste

	MTRU Volume (%)	TRU Volume (%)
Organic materials	53	50
Pu-238	91	72
PCBs	1	—
Reactive	13	—
Corrosive	35	—
Remote handled	4	3
Aqueous, organic, and unique category wastes	0.6	—

Organic material content is one measure of the ability of the waste to generate hydrogen, others being the type and energy of associated radioactivity and possible microbiological activity.[3] A high percentage of both TRU and MTRU wastes contains organic material as well as Pu-238. Significant work is in progress to understand the mechanism of hydrogen formation resulting from radiolysis and methods to control the accumulation of hydrogen during shipment to WIPP. Ongoing research is currently focused on developing materials that will absorb ("getter") the hydrogen. Large volumes of air are circulated throughout the underground disposal areas at WIPP, and generation of hydrogen after placement at WIPP is not regarded as a problem.

As noted in Table B.2, confidence in the chemical composition of the MTRU waste is very low, 16 percent. The reason for the designation of 13 percent of the MTRU volume as reactive and 35 percent as corrosive is unclear. Direct examination and analysis are needed to make a positive determination of these parameters as well as to determine which containers may exceed headspace flammability limits or contain free liquids. In addition, waste characterization to confirm "acceptable knowledge" is required to quantify waste for WIPP disposal (see Sidebar 3.1).

[3]The database does not identify organic material content directly other than those materials associated with RCRA hazardous wastes (toxic organics and hazardous solvents). However, the text description of the individual waste streams makes frequent mention of combustible organics (e.g., paper, wood, cloth, plastics) that are not RCRA materials but are subject to radiolysis and the generation of hydrogen. The volume of waste containing these materials has been estimated by searching for these references.

C

History of Alternatives to Incineration

Incineration has been a tool for managing wastes containing low levels of radioactive contamination since the early stages of the nuclear industry. A review published almost 30 years ago described radioactive waste incineration in the United States and eight foreign countries (Perkins, 1976). In 1992, WASTECH, a multi-organization cooperative project managed by the American Academy of Environmental Engineers with grant assistance from the Environmental Protection Agency (EPA), Department of Defense (DOD), and Department of Energy (DOE), conducted a two-year study resulting in an eight-volume monograph series on innovative site remediation technologies. One volume was devoted to thermal processes, including a variety of incinerator designs and alternatives to incineration.

Recent public opposition to incinerators has forced DOE and other organizations to investigate alternative technologies. As a result, the DOE and several advisory groups have reviewed various technologies and written numerous reports. In April 1997, the Idaho National Engineering and Environmental Laboratory (INEEL) published a detailed report *Evaluation of Alternative Nonflame Technologies for Destruction of Hazardous Organic Wastes* (Schwinkendorf et al., 1997). The report evaluated technologies that are alternatives to open-flame, free-oxygen combustion. Alternative technologies were defined as those that have the potential to destroy organic materials without use of open-flame reactions with free gas-phase oxygen as the reaction mechanism, reduce the off-gas volume and associated contaminants emitted under normal operating conditions, eliminate or reduce the production of dioxins and furans, and reduce the potential for excursions in the process that can lead to accidental release of harmful levels of chemical or radioactive materials.

The report identified 23 technologies and rated them for performance; readiness for deployment; and environmental, safety, and health risks.

The top 10 technologies resulting from this evaluation are

1. steam reforming,
2. electron beam oxidation,
3. ultraviolet (UV) photo-oxidation,
4. ultrasonic destruction,
5. Eco Logic (hydrogen) reduction,
6. supercriticial water oxidation,
7. cerium mediated electrochemical oxidation,
8. DETOX (iron-catalyzed, low-temperature oxidation),
9. direct chemical oxidation, and
10. neutralization or hydrolysis.

The study recommended continuing research to improve incineration and other thermal systems, including air pollution control systems and continuous air emission monitors because *none of the evaluated alternative technologies alone has the capability of thermal systems to treat the large variety of mixed low-level waste (MLLW) in the DOE complex in a single process.* In addition, all of the evaluated alternative technologies have difficulty in treating organically contaminated inorganic matrices such as soils, inorganic sludges, and debris.

In 1997-1998, DOE's Mixed Waste Focus Area published a number of Innovative Technology Summary Reports on acid digestion of organic waste and direct chemical oxidation. More recently, the Transuranic and Mixed Waste Focus Area published brief descriptions of several technologies including mediated electrochemical oxidation, plasma arc systems and direct-current (DC) arc melters, reverse polymerization, solvated electron dehalogenation, steam reforming, and supercritical water oxidation.[1]

In April 2000, following a dispute over the proposed construction of an incinerator for treatment of radioactive mixed waste at INEEL, DOE appointed a blue-ribbon panel of independent experts to explore alternatives to incineration that might become available for use at DOE facilities nationwide (DOE, 2000b). The panel evaluated technologies in five general categories: (1) thermal treatment without incineration; (2) aqueous-based chemical oxidation; (3) dehalogenation; (4) separation (soil washing, solvent extraction, and thermal desorption); and (5) biological treatment. Among the alternatives, the panel considered the most promising to be

---

[1]See http://tmfa.inel.gov/newpages/TechDocs.asp?category=Alternatives%20to%20Incineration).

- thermal (vacuum) desorption of polychlorinated biphenyls, hydrocarbons, and water;
- direct steam reforming to destroy or remove problem contaminants; and
- DC arc and plasma torch melters to destroy contaminants.

The panel found that while there are promising technological alternatives to incineration, none of the alternatives is ready for immediate implementation; all need to be further developed, adapted, and tested with actual waste. The panel therefore recommended a DOE program to demonstrate commercial technologies, nurture the development of next-generation technologies, and guide basic and applied research for future technical advances.

In January 2001, the Secretary of Energy accepted the recommendations of the panel and directed the Office of the Assistant Secretary for Environmental Management to develop an action plan. *Action Plan for Emerging Technological Alternatives to Incineration* was published in June 2001 (DOE, 2001c). The Alternatives to Incineration Committee (ATIC) was formed to examine emerging alternatives and interface with concerned citizens (see Chapter 2, Sidebar 2.4).

The committee reviewed the Action Plan and noted that DOE sites have been largely successful in obtaining relief from regulations that restricted shipment of untreated wastes to the Waste Isolation Pilot Plant (WIPP), which was a principal strategy described in the Action Plan. For example, following a recent revision of the safety analysis for the TRUPACT-II shipping container, the amount of TRU waste that cannot be shipped due to potential hydrogen generation from untreated (nonincinerated) organics is only about 2 percent of the TRU waste inventory (Curl et al., 2002). DOE sites have also been successful in finding alternatives to incineration for most other special case wastes, for example, stabilizing organic liquids on polymers or clays. Because its emphasis is on sending TRU wastes to WIPP, DOE perceives few current incentives to develop true replacement technologies for incineration—those that would destroy essentially all organic materials in a wide variety of wastes and provide large volume reductions of combustible wastes. Developing these technologies, which may be required to treat large volumes of MLLW to meet EPA disposal requirements, remains a challenge as discussed in Chapter 3.

D

Biographical Sketches of Committee Members

DUSCHA, LLOYD A., CHAIR

Lloyd A. Duscha (NAE) is a consulting engineer whose experience encompasses environmental restoration, policy development, organizational management, project management, water resource planning, design and construction, and formulating better contracting practices. He has more than 40 years of experience, including 20 years in executive management positions with the U.S. Army Corps of Engineers culminating as the ranking civilian: deputy director of engineering and construction. Concurrently, he served on the Research and Development Board. Mr. Duscha earned his bachelor's degree in civil engineering, with distinction, from the University of Minnesota, where he was awarded the Board of Regent's Outstanding Achievement Award. Mr. Duscha was elected to the National Academy of Engineering (NAE) in 1987. He is a fellow of the American Society of Civil Engineers and the Society of American Military Engineers. He has served on numerous committees at the National Academies including the Board on Infrastructure and the Constructed Environment (1994-1997); the Committee to Assess the Policies and Practices of the Department of Energy (DOE) to Design, Manage, and Procure Environmental Restoration, Waste Management, and Other Construction Projects (1998-1999); and principal investigator for the Project on Assessing the Need for Independent Review of DOE Projects (1997-1998). Mr. Duscha is currently serving on the National Research Council (NRC) Committee to Review and Assess DOE Project Management (2000-2003).

BURNS, CAROL J.

Carol J. Burns is the deputy division leader of the Chemistry Division at Los Alamos National Laboratory. Dr. Burns is responsible for technical, administrative, and operational management of basic and applied research activities in chemical synthesis and processing, radionuclide and nuclear chemistry, chemical dynamics, instrumentation and diagnostics, and analytical chemistry. She maintains an active research program in actinide and technetium chemistry. Her more than 14 years of service to Los Alamos National Laboratory have included work as deputy group leader, Chemical and Environmental Research and Development Group (1997-1999); program manager for Advanced Concepts, PDET (Energy Technology Programs Office) (1994-1997); and team leader for Inorganic Chemistry, CST-3 (Structural and Inorganic Chemistry) (1991-1994). Dr. Burns is a member of the American Association for the Advancement of Science, the American Chemical Society, Sigma Xi, and the DOE Office of Basic Energy Sciences Council on Chemical Sciences. Her awards include the International Women's Forum Leadership Foundation Fellowship (1998), the National Performance Review Hammer Award (1996), and participation in the National Academy of Sciences' First Annual Symposium on the Frontiers of Science (1989). Dr. Burns has coauthored about 80 peer-reviewed papers in actinide, lanthanide, and technetium inorganic and organometallic chemistry. She earned her B.A. degree in chemistry at Rice University in 1983 and her Ph.D. in inorganic chemistry from the University of California, Berkeley, in 1987 as a Hertz Foundation fellow. She was a J. Robert Oppenheimer postdoctoral fellow at Los Alamos from 1987 to 1989.

COLTON, RICHARD J.

Richard J. Colton is the supervisory research chemist and head of the Surface Chemistry Branch at the Naval Research Laboratory (NRL). He manages a research program in surface chemistry and physics. Program areas include chemical and biological sensors for single-molecule detection, surface science, nanometer-scale science and technology, chemical dynamics, tribology, and coatings. Dr. Colton worked as research chemist and head of the Advanced Surface and Spectroscopy Section of the Surface Chemistry Branch (1982-1998) where he directed research programs on surface and materials characterization by electron spectroscopy and secondary ion mass spectrometry; the study of surface and molecular adsorbate structure using scanning tunneling microscopy; the measurement of adhesive, frictional, and mechanical properties of surfaces using atomic force microscopy; and the development of novel physical, chemical, and biological sensors using electron tunneling and molecular recognition. Dr. Colton has several achievements and honors such as the Federal Laboratory Consortium Award for Excellence in Technology Transfer (2001); Sigma Xi Applied Science Award (1999); 31st Edison Patent Award (1999); 1996 and 2000 NRL Technology Transfer Awards; and the Hillebrand Prize, Chemical Society of Washington (1992). He is a fellow of the American Vacuum Society and is currently serving on the NRC Board on Assessment of NIST (National

Institute of Standards and Technology) Programs, Subpanel for JILA (formerly Joint Institute for Laboratory Astrophysics). Dr. Colton has authored or coauthored more than 130 articles and book chapters in scientific journals and monographs. Dr. Colton earned his B.S. degree in chemistry with a minor in mathematics in 1972 and his Ph.D. in physical chemistry in 1976 from the University of Pittsburgh.

KEARFOTT, KIMBERLEE J.

Kimberlee J. Kearfott is a professor in the Department of Nuclear Engineering and Radiological Sciences, with a joint appointment in the Department of Biomedical Engineering, at the University of Michigan, Ann Arbor. Her fields of expertise include radiation detection, medical and tomographic imaging, nuclear medicine and diagnostic radiology physics, external and internal dosimetry, medical and nuclear power plant health physics, and physiological models. Dr. Kearfott has authored or coauthored more than 250 publications, including over 65 full-length papers in peer-reviewed journals. She has delivered more than 150 conference presentations, and 115 guest lectures, and has made 9 radio and television appearances. Dr. Kearfott has several achievements and honors including the Women's Achievement Award from the American Nuclear Society (1995); the Elda Anderson Award from the Health Physics Society (1992); the Society of Nuclear Medicine Tetalman Award (1991); and the National Science Foundation Presidential Young Investigator Award (1985). She is a member of the American Association of Physicists in Medicine, the American Nuclear Society, the Society of Nuclear Medicine, Sigma Xi, the Association of Women in Science, the Institute of Electrical and Electronics Engineers, the Society of Women Engineers, the International Radiation Physics Society, and the Order of the Engineer. Dr. Kearfott has been a radiological engineer for Detroit Edison Fermi I and Fermi II Nuclear Power Facilities, an associate professor at the Georgia Institute of Technology and the Arizona State University, and a research associate for the Sloan-Kettering Institute for Cancer Research. Dr. Kearfott earned her B.S. degree, diploma in engineering, from St. Mary's University, Nova Scotia, in 1975; her M.E. in nuclear engineering from the University of Virginia in 1977; and an Sc.D. in nuclear engineering from the Massachusetts Institute of Technology with a doctoral minor in physiology and medical physics in 1980.

SAMELSON, RICHARD J.

Richard J. Samelson retired in 1994 from PPG Industries, Pittsburgh, Pennsylvania, after 40 years of service. While at PPG, he worked within the Chemicals Group as manager and director of Environmental Programs, manager of Technical Support Systems, manufacturing engineer for Inorganic Chemicals, process engineer at the Natural Soda Ash

Facility in Bartlett, California, and R&D engineer at Corpus Christi, Texas. Mr. Samelson's later responsibilities included environmental management and control, risk evaluation, and management of projects for the investigation and control of air and water pollution associated with past waste disposal practices. During his career, he served as a member of the Chemical Manufacturers Association Environmental Management Committee and the Environmental Protection Committee of the Chlorine Institute, where he served two years as vice chairman and chairman, respectively. Mr. Samelson served on the NRC Committee on Mixed Wastes. He earned a B.S. degree in chemical engineering from Iowa State University in 1954.

STEFFAN, ROBERT J.

Robert J. Steffan is the vice president of Technology Development for Envirogen, Inc. Dr. Steffan's areas of expertise include in situ bioremediation, fermentation technologies, and advanced technologies of biotransformations, gene probes, genetic engineering, and novel treatment methods. He has lectured widely on the topics of development of biocatalysts, use of molecular biology in hazardous waste treatment, and biodegradation. Dr. Steffan has coauthored 45 articles on his areas of expertise in monographs and scientific journals. He has served on the editorial board of the *Journal of Bacteriology,* and currently serves on the editorial board of *Applied and Environmental Microbiology.* He holds several patents including a method to treat toxic chemicals and another to degrade and remediate organochlorides. Dr. Steffan's career at Envirogen, Inc., has been as director, Bioremediation and Advanced Technologies Research (1998-2001); research manager, Bioremediation Technologies (1994-1998); manager, Genetic Engineering Group (1993-1994); and research scientist (1990-1993). He was a research scientist (1989-1990) and an Alexander von Humboldt-Stiftung research fellow (1988-1989) in the Department of Microbiology at the Gesellschaft für Biotechnologische Forschung, Braunschweig, Germany. Dr. Steffan's honors and achievements include the Thomas Alva Edison Patent Award in 2000. Dr. Steffan earned an associate of arts degree from Bismarck Junior College, North Dakota in 1979; a B.A. in Biology from Jamestown College, North Dakota, in 1982; an M.S. in biology from the University of Wisconsin, LaCrosse, in 1984; a Ph.D. in biology from the University of Louisville in 1988; and a J.D. from Temple University School of Law in 1997.

TSCHINKEL, VICTORIA J.

Victoria J. Tschinkel consults in environmental policy and planning in Tallahassee, Florida. Her expertise is in assisting corporate clients on strategic environmental issues and in representing clients before agencies and the state legislature. Ms. Tschinkel is a director of Phillips Petroleum Company, Resources for the Future, and the Center for Clean Air Quality. She is a member of the National Academy of Public Administration. Ms. Tschinkel is an Environmental Regulation Commissioner for the State of Florida. She served as the secretary of the Florida Department of Environmental Regulation (1981-1987) and has held positions on a number of national advisory councils such as the National Environmental Enforcement Council and the Energy Research Advisory Board. She currently serves as a member of the NRC's Board on Radioactive Waste Management and is a former member of the Commission on Geosciences, Environment, and Resources. She has served on numerous NRC study committees, including the Committee to Evaluate the Science, Engineering, and Health Basis of the Department of Energy's Environmental Management Program, the Committee on Remedial Action Priorities for Hazardous Waste Sites; and the Committee to Provide Interim Oversight of the DOE Nuclear Weapons Complex. Ms. Tschinkel earned her B.S. degree in zoology from the University of California, Berkeley.

UHLE, MARIA E.

Maria E. Uhle is the Jones Assistant Professor of Environmental Organic Geochemistry in the Department of Geological Sciences at the University of Tennessee. Her research includes investigating the organic chemical composition of atmospheric particulates; the influence of dissolved organic material (DOM) on the fate of organic pollutants in aquatic environments, and how contaminants bind to DOM. Dr. Uhle has been a postdoctoral fellow at the School of Environmental and Marine Sciences, the University of Auckland (1999-2000), and the Department of Geological Sciences and Byrd Polar Research Center at Ohio State University (1998). She has coauthored several publications in scientific journals. Dr. Uhle earned her B.S. in geology from Bates College in Lewiston, Maine, in 1988; her M.S. in geology from the University of Massachusetts, Amherst, in 1992; and her Ph.D. in environmental sciences from the University of Virginia, Charlottesville, in 1997.

ZYLSTRA, GERBEN J.

Gerben J. Zylstra is a professor in the Biotechnology Center for Agriculture and the Environment, director of the Nucleotide Sequencing Facility, and director of the High Throughput Screening Laboratory at Rutgers University. His areas of expertise are microbiology, genetics, and biochemistry of the degradation of hydrocarbons. Dr. Zylstra has collaborated with experts from many different fields in numerous publications. He has been invited to give seminars at local, national, and international meetings in the United States, Egypt, Israel, South Korea, Mexico, Puerto Rico, Venezuela, Brazil, Taiwan, Germany, the Netherlands, and Spain. His honors and awards include election to the American Academy of Microbiology in 2001, the Selman A. Waksman Award in 2001, the Cook College and New Jersey Agricultural Experiment Station Sustained Research Excellence Award in 2000, the Foundation for Microbiology lecturer award for 1997-1999, and the National Science Foundation Young Investigator Award for 1992-1997. Dr. Zylstra was a postdoctoral research associate in the Department of Microbiology at the University of Iowa, Iowa City (1988-1990), and the University of Texas, Austin (1987-1988). Dr. Zylstra is an editor of *Archives of Microbiology* and is on the editorial board of the *Journal of Microbiology.* He is a member of numerous groups at Rutgers, including the Ocean Science Engineering Center and the Deep Sea Ecology and Biotechnology Center, and is chair of the Agricultural and Environmental Genomics Committee. Dr. Zylstra earned his B.S. degree in biology in 1981 at Calvin College, Grand Rapids, Michigan, and his Ph.D. in cellular and molecular biology in 1987 at the University of Michigan Medical School.

E

Presentations to the Committee

Washington, D.C., May 31-June 1, 2001

Overview of the Environmental Management Cleanup Mission and the Office of Science and Technology, Gerald Boyd, Department of Energy (DOE) Office of Environmental Management (DOE-EM)

Needs and Opportunities for Transuranic (TRU) and Mixed Waste Research, Mark Gilbertson, DOE-EM

The Role of the Office of Science in the Environmental Management Science Program, Roland Hirsch, DOE Office of Biological and Environmental Research

TRU Waste Management, Douglas Tonkay, DOE-EM

EM's Mixed Low-Level Waste Management Program, Helen Belencan, DOE-EM

TRU and Mixed Waste Focus Area Overview, Edward Rizkalla, DOE-EM

DOE Response to the Secretary of Energy Advisory Board Panel Report on Technological Alternatives to Incineration, Helen Belencan, DOE-EM

Idaho Falls, Idaho, August 6-7, 2001

Overview of the Idaho National Engineering and Environmental Laboratory (INEEL), Lisa Green, DOE-Idaho

Characteristics of INEEL Stored TRU Waste, Tom Clements, Jr., INEEL

TRU Waste Treatment and Disposal Plans, Tom Clements, Jr., INEEL

Science and Technology for Characterizing, Treating, and Disposing of Mixed and TRU Waste, Michael Connolly, INEEL

Advanced Mixed Waste Treatment Project, Fred Hughes, BNFL, Inc.

Waste Generator Services Mixed Waste Project, Jeffrey Mousseau, INEEL

Oak Ridge, Tennessee, September 18-19, 2001

Oak Ridge TRU Waste Management, Gary Riner, DOE-Oak Ridge

Mixed Low-Level Waste Program, Fred Heacker, Bechtel Jacobs Corp.

Overview of Technology Demonstrations for Monitoring Emissions from the TSCA[1] Incinerator, J. E. Dunn, Jr., IT Corp.

Waste Issues from the Spallation Neutron Source, Frank Kornegay, Oak Ridge National Laboratory

Aiken, South Carolina, September 20, 2001

Savannah River Site and Waste Management Operations, Jim Blankenhorn, Westinghouse Savannah River Company

Development of a Microbial Process for Removal of Organic Constituents from PUREX[2] Waste, Michael Heitkemp, Savannah River Technology Center (SRTC)

Testing Hydrogen Getters for TRUPACT-II Payload Expansion, Jon Duffey, SRTC

Pu-238 Decontamination Demonstration—Silver II, Bob Pierce, SRTC

---

[1]Toxic Substance Control Act.
[2]Plutonium and Uranium Extraction.

PUREX Waste Alternative Treatment Evaluation, Marshall Looper, SRTC

PUREX Waste Stabilization—Nochar and Imbiber Bead Polymers, Christine Langton, SRTC

Pretreatment of Legacy PUREX Waste, Major Thompson, SRTC

Carlsbad, New Mexico, December 5, 2001

National TRU Waste Program Vision for the Future, Inés Triay, DOE-Carlsbad

Recent National Research Council Recommendations for the Waste Isolation Pilot Plant, Roger Nelson, DOE-Carlsbad

Optimization and TRU Technology Needs, Dave Moody, Los Alamos National Laboratory-Carlsbad Operations

Treatment of TRU Waste, Robert Behrens, Los Alamos National Laboratory-Carlsbad Operations

Richland, Washington, February 12, 2002

Hanford Waste Management Program, Dale McKenney, Fluor Hanford

Plans for Treating Remote-Handled TRU Waste at T-Plant, Bob Barmettlor, Fluor Hanford

Hanford Mixed Low-Level and TRU Waste Management Needs, Kevin Leary, DOE-Richland, and Wayne Ross, Pacific Northwest National Laboratory (PNNL)

Large Contaminated Equipment Project, Betty Carteret, PNNL

Nondestructive Waste Assay, Tony Peurrung, PNNL

Research and Development for a Remotely Operated, Multipurpose Robotic Vehicle, Kevin Leary, DOE-RL

In addition to the above presentations, the committee participated in informative roundtable discussions with site research scientists, as follows:

Oak Ridge National Laboratory, September 19, 2001
Savannah River Technology Center, September 20, 2001
Los Alamos National Laboratory, December 6, 2001
Sandia National Laboratory, December 6, 2001

The committee also heard presentations by citizen groups at INEEL, the Oak Ridge Reservation, and the Savannah River Site.

F

List of Acronyms

ATIC	Alternatives to Incineration Committee
CERCLA	Comprehensive, Environmental Response, Compensation, and Liability Act (Superfund)
CH-TRU	contact-handled TRU waste
DGGE	density gradient gel electrophoresis
DOD	Department of Defense
DOE	Department of Energy
DOT	Department of Transportation
EM	DOE Office of Environmental Management
EMSP	Environmental Management Science Program
EPA	U.S. Environmental Protection Agency
FFCA	Federal Facility Compliance Act of 1992
HEPA	high-efficiency particulate arresting
HWC	hazardous waste combustor
HWIR	Hazardous Waste Identification Rule
INEEL	Idaho National Engineering and Environmental Laboratory
LDR	Land Disposal Restriction
LLW	low-level radioactive waste
MACT	Maximum Achievable Control Technology
MEMS	MicroElectroMechanical Systems
MLLW	mixed low-level waste
MOU	Memorandum of Understanding
MRI	magnetic resonance imaging

MTADS	Multi-sensor Towed Array Detection System
MTRU	mixed transuranic waste
MWIR	Mixed Waste Inventory Report
NABIR	Natural and Accelerated Bioremediation Research
NESHAP	National Emission Standards for Hazardous Air Pollutants
NDA	nondestructive assay
NDE	nondestructive examination
NEMS	NanoElectroMechanical Systems
NRC	National Research Council
NRL	Naval Research Laboratory
OST	DOE Office of Science and Technology
PCB	polychlorinated biphenyl
PCR	polymerase chain reaction
RCRA	Resource Conservation and Recovery Act of 1976
RH-TRU	remote handled TRU waste
RTR	real-time radiography
RWMC	Radioactive Waste Management Complex at Hanford
SAMS	Surveillance and Measurement System
SARA	Superfund Amendments and Reauthorization Act
SARP	Safety Analysis Report for Packaging
SAW	surface acoustic wave
TCLP	toxicity characteristic leaching procedure
TM	transuranic and mixed waste
TMFA	Transuranic and Mixed Waste Focus Area
T-RFLP	terminal restriction fragment length polymorphism
TRU	transuranic
TSCA	Toxic Substances Control Act
USNRC	U.S. Nuclear Regulatory Commission
VOC	volatile organic compound
WAC	waste acceptance criteria
WETO	Western Environmental Treatment Office
WIPP	Waste Isolation Pilot Plant
the Permit	WIPP Waste Acceptance Permit